MARTIN FRANCIS

DESIGN & INNOVATION

MARCUS KRALL

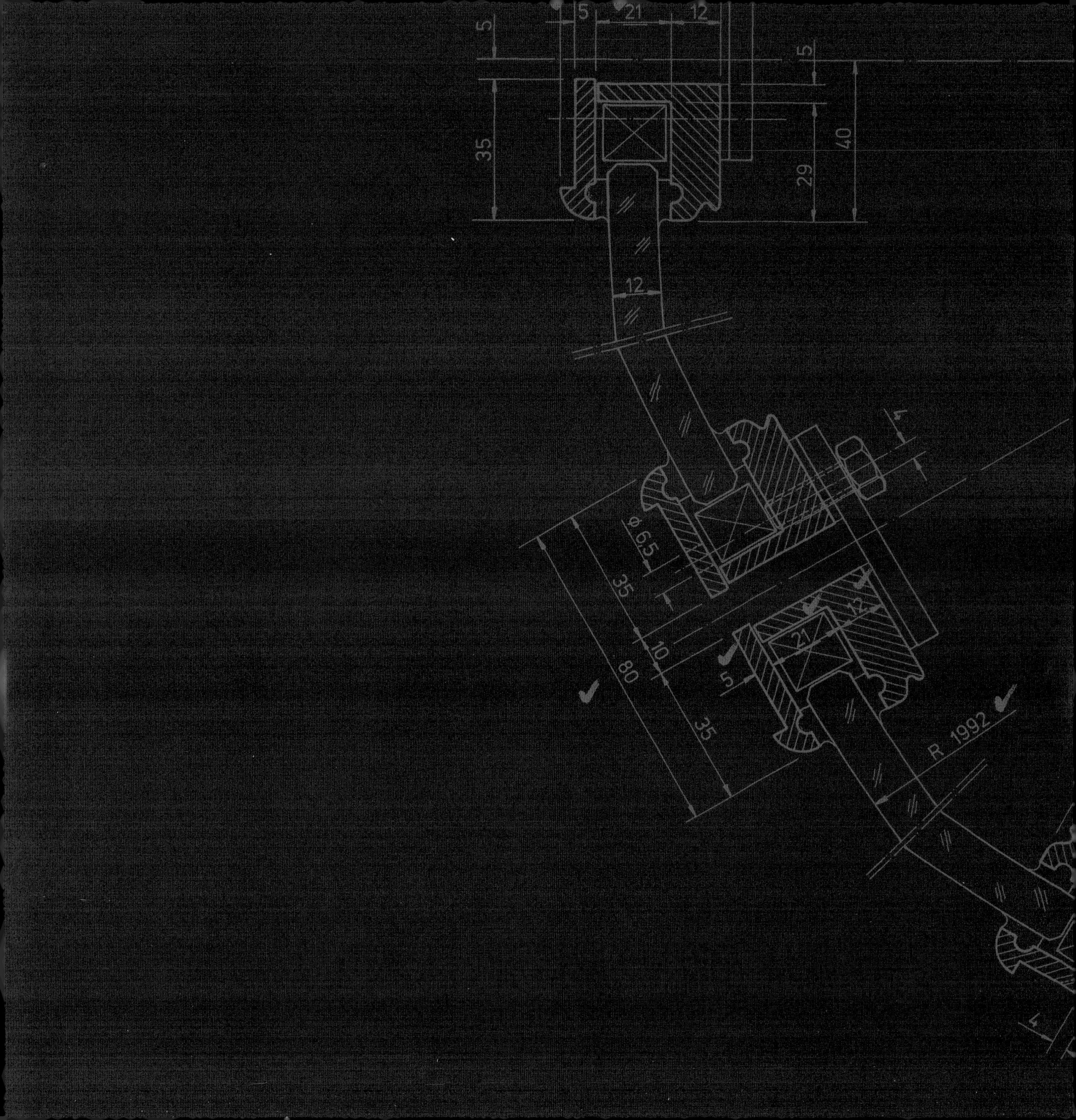

5
21
12
5
35
29
40
12
Φ 6,5
35
10
80
21
12
5
35
R 1992

MARCUS KRALL

MARTIN FRANCIS

DESIGN & INNOVATION

Koehler

Inhalt Content

Vorwort
Preface

Als ich noch ein Kind war, besuchten meine Eltern mit mir häufiger das Wissenschaftsmuseum in London. Ich erinnere mich gern daran, denn dort entdeckte ich Brunel, der bis heute eines meiner großen Vorbilder ist. Er war ein Mann, der einen Zug entwarf, um die Menschen von London nach Bristol zu bringen, dann die Docks in Bristol und dann das Schiff, um sie nach New York zu bringen. Dieser Brunel war der Elon Musk seiner Zeit, eine ständige Quelle der Inspiration. Unter anderem hatte das Museum früher auch einen wunderschönen Bereich über den Schiffbau, aber zu meiner Bestürzung ist er heute zu einer Mischung aus Disney World und Einkaufspassage verkommen.

Der Grund, warum ich das alles erwähne, ist, dass Marcus Krall mich mit Peter Tamm bekannt gemacht hat, der mich zu einem Besuch seines Internationalen Maritimen Museums in Hamburg und seines Archivs mitnahm – ein unglaubliches Vergnügen für mich, denn dieses Museum ist eines, wie es sein soll. Als ich es verließ, rief ich sofort meine Frau an und sagte: »Wenn ich auf dem Heimweg von einem Bus angefahren werde, schick bitte mein ganzes Archiv hierher.«

So begann die Zusammenarbeit zwischen Martin Francis, Marcus Krall und Peter Tamm, die in der Produktion dieses Buches gemündet ist. Ich bin sehr dankbar dafür, denn ich habe zeit meines Lebens eine Schreibschwäche gehabt und es immer vermieden, viel zu publizieren. Durch die Unterstützung der richtigen Menschen ist dieses Buch das erste, das meine bis heute 55-jährige Karriere reflektiert.

Der gemeinsame Nenner bei allem, was ich getan habe, waren die bemerkenswerten Menschen, mit denen ich das Privileg hatte zusammenzuarbeiten. Ob Kunden oder Kollegen, sie waren für die Entwicklung meiner Karriere von grundlegender Bedeutung.

Gute Projekte entstehen aus guten Kunden – Menschen, die einen herausfordern, aber ihrerseits gerne ihre Vorurteile hinterfragen lassen und vor allem Spaß am Prozess haben.

When I was a child, I frequently visited the Science Museum in London, it was there that I discovered Brunel one of my great idols. He was a man who designed and built a train to take people from London to Bristol, the docks in Bristol and then the ship to take them to New York, the Elon Musk of his day, a constant source of inspiration. Among other things the museum used to have a wonderful section on naval architecture but much to my dismay it has now become a cross between Disney world and a shopping arcade.
The reason I mention all this is that Marcus Krall introduced me to Peter Tamm who took me to visit his International Maritime Museum in Hamburg and his archives, what a pleasure, this is what a museum should be. Upon leaving the museum, I called my wife and said, "If I get hit by a bus on my way home, this is where to send my archives".
Thus, started the collaboration that has resulted in the production of this book, being dyslectic I have always had a problem with writing, indeed until the advent of word processing it was impossible, I have therefore managed to avoid expressing myself in print for all my life and this is no exception. Happily, Marcus has undertaken this brief glimpse into some of the projects that I have been lucky enough to undertake over the last 55 years.
The common thread with everything I have done has been the remarkable people I have had the privilege of working with. Be they clients or my peers, they have been fundamental to the development of my career.
Good projects come out of good clients, people who challenge one but who in turn like to have their preconceptions challenged and above all who enjoy the process.

Interview
Interview

Das Navigationsgerät hat aufgegeben. »À droite, à gauche« plärrt es, und dabei geht es weder nach links noch nach rechts, sondern nur ziemlich steil den Berg hinauf. Mein Gesprächspartner hatte mich vorgewarnt: Allradantrieb wäre wichtig – »check« – und ein paar Brocken Französisch, wenn ich den Weg nicht finden würde und Nachbarn fragen müsste. Dies ist nun der Fall, und die älteren Herren deuten zurück, grübeln kurz, schicken mich schließlich nach links, rechts, links und dann nach oben.

Als ich dann wie vereinbart an der Einfahrt hupe, schmunzelt Martin Francis zum Empfang. »Gut versteckt, oder?« Wir befinden uns in Grasse, also nicht ganz am Meer, aber doch in Sichtweite und nahe genug an den Yachting-Hotspots der Côte d'Azur.

Schon im Carport gibt es einen Vorgeschmack darauf, was mich gleich erwartet. Unter der Decke und unmittelbar über meinem Auto hängt das Tanktest-Modell der Motoryacht A. Für das ungewöhnliche 119-Meter-Format von Andrei Melnitschenko hatte Francis die Konstruktion und das technische Design geliefert – in der Projektbezeichnung SF99 steht das SF für Starck/Francis. Doch wir wollen heute über viel mehr sprechen als über die Ablieferungen der jüngeren Zeit. Woher kommt einer der einflussreichsten Yachtdesigner unserer Zeit überhaupt? Wie hängen seine Aktivitäten im Yachting, in der Architektur und der Kunst zusammen?

Für das Gespräch führt er in die moderne Villa, die sich perfekt in die Landschaft einfügt. Gebaut beziehungsweise konstruiert hat er das Gebäude nicht, aber sich sofort bei der Besichtigung darin verliebt. »Ich musste es einfach haben.« Drinnen stehen allerlei Devotionalien von vielen Reisen, etliche Bücher, in seinen Arbeitsbereich, einen flachen Anbau, darf ich nur kurz schauen – die Unordnung ist ihm etwas unangenehm, »kreatives Chaos halt«.

The navigation system has given up. It's calling out "à droite, à gauche" but there's no way to go left or right, the only option is straight up a very steep hill. My potential interviewee had warned me: I'd need four-wheel drive – check – and a bit of French in case I got lost and had to ask the neighbours. This is just what I've had to do and the old man I asked pointed back the way I came, reconsiders, and, finally, sends me left, right, left, and then up.

When I beep my horn at the entrance gate as agreed, Martin Francis smiles in greeting. "Pretty well hidden, isn't it?" We are in the hills above the so-called capital of perfume, Grasse, not right on the ocean, but in sight of the sea and close enough to the yachting hotspots of the Cote d'Azur.

The carport provides a taste of what awaits me. The tank test model of the motor yacht A is hanging from the ceiling just above my car. Francis delivered the construction and technical design for Andrey Melnichenko's unusual 119 metre first motor yacht – the SF in the project name SF99 stands for Starck/Francis. But we have a great deal more to talk about today than just the most recent deliveries. Where does one of today's most influential yacht designers come from? What inspires him? How are his activities in yachting, architecture, and art related to one another?

For the interview, he leads me into his modern villa, which fits in perfectly with the landscape. He didn't build, or rather design, the house, but he did fall in love with it immediately when he saw it. "I just had to have it." Inside, the house is full of memorabilia from his travels and countless books; I'm only permitted a quick glance into his studio, a low add-on – he's somewhat embarrassed by the mess, "it's creative chaos".

▸

**Martin, Sie gelten als einer der einflussreichsten Yachtdesigner unserer Zeit.
Viele Ihrer heutigen Kollegen oder Mitbewerber haben bei Ihnen das Handwerk gelernt.
Wie sind Sie eigentlich gestartet?**
Ja, so einige meiner Schützlinge von damals wie etwa Espen Øino, Dan Lenard, Mark Smith oder Jonny Horsfield haben tolle Karrieren hingelegt und ein schönes Portfolio geschaffen. Sie hatten auch ein bisschen Glück, dass das Yachtdesign schon etwas entwickelter war als zu der Zeit, als ich damit begann. Ich entdeckte diese Disziplin erst relativ spät, zumindest nicht direkt nach oder auf der Hochschule.

**Nur wenige Menschen wissen wahrscheinlich, wie Ihr Berufsleben begonnen hat.
Mögen Sie es kurz skizzieren?**
Das ist ziemlich lange Geschichte.

Ich fühle mich ganz wohl hier.
Ok (schmunzelt). Eine recht wichtige Voraussetzung war wahrscheinlich, dass ich aus einem recht kreativen Elternhaus komme. Meine Mutter hatte am Royal College of Art und an der Chelsea School of Art bei Henry Moore studiert, mein Vater war neben seiner Ausbildung als Arzt ein erfolgreicher Industrie- und Architekturfotograf. Er war zudem ein begeisterter Handwerker und baute seine eigenen Kameras, während meine Mutter Spielzeuge für meinen Bruder und mich im Arbeitszimmer der Familie fertigte – es war die Zeit kurz nach dem Krieg, und es gab nicht viel. Das Familienmotto hieß deshalb: Wenn du etwas möchtest, bau es dir. Diese frühen Erlebnisse, dass aus einigen wenigen und ganz einfachen Materialien ein Produkt entsteht, hat mich geprägt. Ein ganz entscheidender Grundsatz von mir ist, dass man nicht Dinge zeichnen kann, von denen man nicht weiß, wie sie gebaut werden.

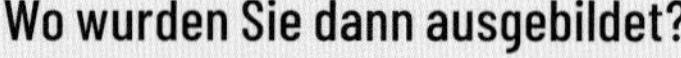

Wo wurden Sie dann ausgebildet?
Mit 13 Jahren kam ich auf ein Internat, weil bei mir eine Dyslexie vermutet wurde und ich nicht so ganz in das standardisierte Schulsystem der damaligen Zeit passte. Auf dem Internat bekam ich mehr Freiheiten und durfte in der Werkstatt meiner Design- und Bastelleidenschaft,

**Martin, you are considered to be one of the most influential yacht designers of our time.
Many of your current colleagues or competitors learned the trade from you.
So how did you get started?**
That's true. Some of my earlier protégés, like Espen Øino, Dan Lenard, Mark Smith and Jonny Horsfield, went on to have great careers and create wonderful portfolios. They were lucky; yacht design is more advanced than when I started out. I only discovered this discipline relatively late in life, or, at least, not right after university or during my studies.

As only a few readers know how your career got started. Would you mind giving us a short summary?
It's a long story.

I feel pretty good here.
Okay (grins). One pretty important factor was probably that I come from a fairly creative family. My mother studied at the Royal College of Art and the Chelsea School of Art under Henry Moore; my father was in addition to being a doctor, a successful industrial and architectural photographer. He was also a passionate handyman and built his own cameras, while my mother made toys, it was just after the war and they weren't available, for my brother and me. The family motto was if you needed something then make it, so experiencing early on how something can be made from simple materials really had an impact on me. One of my basic principles is that you can't design things if you don't know how to build them.

Where were you educated?
When I was 13, I was sent to boarding school because, being dyslectic before it was recognised, I didn't really fit into the standardised school system of the time. At boarding school, I had a lot more freedom and was given the run of the workshop to develop my passion for designing and handicraft, which was just emerging.

die gerade entstand, freien Lauf lassen. Ich baute Tische und Stühle, experimentierte mit Fiberglas, als dieses Material aufkam, und schaffte den Abschluss als 17-Jähriger gerade so eben. Ein Running Gag lautete: Wenn Francis das Fach Englisch schafft, schafft es jeder. Ein guter Schüler war ich nicht, aber ich war wohl recht talentiert in der Gestaltung. Mein Vater verschaffte mir jedenfalls ein Aufnahmegespräch an der heutigen Central Saint Martins, einer besten Designschulen überhaupt. Normalerweise musste man erst woanders studiert haben, um überhaupt vorsprechen zu dürfen. Um es kurz zu machen: Ich nahm meine Tische und Stühle aus dem Internat mit und wurde genommen. Die meisten Kommilitonen waren sehr viel älter, aber auch sehr inspirierend; zudem kümmerten sie sich rührend um mich. Mit 20 Jahren hatte ich mein Diplom in der Tasche und betrieb mit einem alten Schulkameraden eine Möbeltischlerei.

Davon verstanden Sie ja eine Menge.

Ja, wir hatten mit den Möbeln auch einen gewissen Erfolg, verkauften sie an Shops wie Heal's und direkt an Endkunden. Nach einer gewissen Zeit wurde mir jedoch langweilig. Einer unserer Nachbarn war der Ingenieur Tony Hunt, der mich fragte, ob ich nicht mit ihm zusammenarbeiten wolle. Hunt machte mich dann mit Norman Foster bekannt, mit dem er für einige Projekte kooperierte. Foster war noch unbekannt, das war 1967/68, nach seiner Team-4-Zeit mit Richard Rogers. Wir verstanden uns alle gut, privat wie beruflich, und waren plötzlich ein Team. Mit einem Entwurf für eine Schule wurde das Büro schließlich bekannter. Wir gewannen die Ausschreibung nicht, aber Foster vermarktete den Entwurf sehr gut.

Das bedeutete, dass nun reichlich Anfragen kamen?

Ja, es kamen Anfragen und als Erstes eine extrem ungewöhnliche. Eines Morgens kam ein Mann in Hunts Büro und fragte, ob wir auch eine spezielle Bühnenkonstruktion abliefern könnten. Ich sah mir das an und fuhr schnurstracks zu einem Zulieferer, den ich kannte. Am nächsten Tag präsentierte ich die Lösung, und unser Auftraggeber war extrem beeindruckt. Er hieß Chip Monck und engagierte mich vom Fleck weg. Es war der Tourmanager der Rolling Stones.

I built tables and chairs, experimented with fibreglass which was a brand-new material and barely managed to graduate when I was 17. A running gag was: if Francis can pass English, anyone can. I may not have been a good pupil, but apparently I was quite a talented designer. At any rate, my father got me an interview at, what is now called, Central Saint Martins, one of the best design schools in London. Normally, you had to have studied elsewhere first before you even got an interview. To make a long story short: I took the tables and chairs I had built at boarding school to the interview and was accepted. Most of the other students were much older and more experienced than me so they taught me a lot. By the age of 20, I had my degree and was running a cabinetmaking business with an old schoolmate.

You certainly knew an awful lot about making furniture.

True, we had a certain degree of success with our furniture and sold it to shops like Heal's and direct to customers. After a while, though, I grew bored. One of our neighbours was the young engineer Tony Hunt who invited me to work with him, he then introduced me to Norman Foster because he and Hunt worked on the same projects. Foster had yet to make a name of himself; it was in 1967/68 just after working with Richard Rogers as Team 4. We all got along very well, both personally and professionally, and became a team. One of the projects we did was a competition for a school which, while we didn't win, got Foster more recognition. While, Foster as usual, did a great job of marketing the brand.

That means you received lots of enquiries after that?

Well yes, we had some, one of the first was extremely unusual. One morning, a man came into Hunt's office and asked if we could calculate a demountable proscenium arch for a stage. I had an idea and headed straight to a supplier I knew. The next day, I demonstrated the solution on the pavement outside the client's office, he was impressed and hired me on the spot. He was called Chip Monck and was the tour manager for the Rolling Stones.

►

Oh, mein Gott …

Das dachte ich ebenfalls. Ich wurde über Nacht praktisch einer der Bühnenplaner der vielleicht wichtigsten Band der Welt. Wir reisten mit einem Tross von etwa 40 Leuten durch Europa, und ich organisierte zusammen mit Chip Monck alle Abläufe. Als Mick Jagger gerade »Jumpin' Jack Flash« sang, warf er mir seine Jacke von der Bühne aus zu, am nächsten Tag saßen wir alle zusammen im Flugzeug zum nächsten Gig.

Was ist Ihnen aus dieser Zeit am stärksten im Gedächtnis geblieben?

Nun, vielleicht das Konzert in Stockholm. Es fand in einem der ersten Fußballstadien mit beheiztem Rasen statt, und die Fans durften natürlich nicht rauf. Bei den Vorgruppen gelang das noch. Als die Stones dann auf die Bühne traten, stürmten die Fans den Innenraum. Der Manager des Stadiums brüllte mich an, dass wir dies sofort stoppen müssten. Ich zuckte nur die Schultern. Wie soll man 50.000 Menschen aufhalten?

Haften geblieben ist mir aus der Zusammenarbeit mit Chip Monck, der zudem ein Co-Presenter des Woodstock-Festivals war, auch, dass man mit dem nötigen Willen und dem nötigen Budget eigentlich alles möglich machen kann. Das hat mir später auf manchen Werften geholfen.

Führten Sie auch dieses typische Leben, von dem man immer hört: Sex, Drugs and Rock 'n' Roll?

Gott, bewahre. Natürlich himmelten Groupies die Stones an, und die eine oder andere Party gab es auch. Mein Eindruck von Jagger war jedoch, dass er auf solchen Tourneen bewusst lebte. Diese Auftritte schlauchten ganz schön. Und Keith Richards war zu dieser Zeit ja mit Familie – Anita Pallenberg und Baby Marlon – unterwegs.

Wie bekommt man dann eigentlich die sogenannte Kurve von der Bühne der Stones zur damals recht unbekannten Disziplin des Yachtdesigners?

Ich unterbrach die Tour mit den Stones, weil mein Vater krank wurde. Als ich zurückkam, hatte man zwischenzeitlich gemerkt, dass die Auftritte auch ohne mich funktionierten. Um die Sache ein bisschen abzukürzen: Ich ging zurück nach England und baute mit John Morris, dem anderen Woodstock-Presenter neben Chip Monck,

Oh, my God …

Yeah, that's what I thought. Essentially overnight, I became the assistant production manager for what was possibly the most important band in the world. We travelled through Europe with a group of about 40 people and I worked with Chip Monck to organise everything. During a concert where I was acting as security standing stage right, Mick Jagger singing "Jumpin' Jack Flash", threw me his jacket for safe keeping.

What is your most vivid memory from this period?

Well, maybe the concert in Stockholm. It was in one of the first football stadiums with under-soil heating, and, of course, the fans weren't allowed on the field. That worked out fine for the opening acts. But when the Stones came on stage, the fans stormed the pitch. The stadium manager yelled at me, saying we had to stop them immediately. I just shrugged. How was I supposed to stop 50,000 hysterical fans?

Something I learnt from working with Chip Monck, who had been the co-presenter of Woodstock the year before, was if you have the will and the right budget, just about anything is possible. That helped me out in a few yacht yards later on.

Did you lead that typical life we all hear about: sex, drugs and rock'n'roll?

God help me. Of course the groupies worshipped the Stones and there were a few parties. But my impression of Jagger was that he lived very consciously on those tours. Those concerts, every other day, were really exhausting. Keith Richards had Anita Pallenberg and their baby Marlon, nanny and toys with then all the time.

So how did you get your act together, as they say, and move from the stage with the Stones to the yet unknown discipline of yacht designer?

I stopped touring with the Stones because my father fell ill. By the time I came back, they had noticed that the concerts went just fine without me. To cut a long story short: I went back to England and joined the other presenter of Woodstock John Morris in the creation of the Rainbow Theatre, a Rock and Roll venue. However, after

das Rainbow Theatre auf, eine Rock 'n' Roll-Konzerthalle. Schnell wurde mir klar, dass es nicht meine Berufung war, solch ein Haus zu leiten. Aus Langeweile begann ich, Boote und Yachten zu zeichnen.
Vorerst aber, also vor meiner Yachtdesign-Karriere, engagierte mich der Architekt John Hix für einen Wohnkomplex in Milton Keynes, danach kam ein Anruf eines Freundes, der bei Foster beschäftigt war. Das Büro hatte seinen ersten großen Auftrag, die Büros von Willis Faber & Dumas in Ipswich. Ich sollte bei der Gestaltung einer riesigen Glasfassade helfen und baute mir über die folgenden drei Jahre den Ruf als »Glass Man« auf.
Zwischenzeitlich hatten meine Frau und ich ein 8,50 Meter langes Segelboot von Van de Stadt aufgebaut, mit dem wir einige Regatten gewannen. Vielleicht, weil das Boot kein Interior besaß oder weil wir ganz gut segelten? Als Upgrade kauften wir dann eine Contessa 32.
1975, als es noch weniger Arbeit für Architekten/Designer wie mich gab – es herrschte eine Rezession –, siedelten wir nach Frankreich um, in die Nähe meiner Schwiegereltern. Ich arbeitete als Agent für Contessa, war jedoch – wie heute noch – kein guter Verkäufer, half bei einem Mastenbauer aus und baute so nebenbei ein Boot, mit dem meine Familie und ich um die Welt segeln wollten.

Was war das für ein Boot?

Eine 14-Meter-Slup, die ich Prototype *nannte. Wir bauten sie in Biot ohne Interior und segelten sie vor Südfrankreich ohne Interior und damit mit einem Höllentempo, wenn die Bedingungen stimmten. Das sprach sich irgendwie herum – Superyacht-Medien gab es ja kaum – und plötzlich bekam ich Yachtdesign-Aufträge für eine 25-Meter-Ketsch und zwei 26-Meter-Slups. Das waren damals gewaltige Schiffe. Kurz darauf kamen weitere Anfragen, und ich war praktisch der Gestalter der vier größten Slups der Welt, die zudem zeitgleich um die Welt segelten.*

Mir war nicht wirklich bekannt, dass Ihre Karriere mit Segelyachten begann.

Das wird heutzutage gern vergessen. Eco *oder* Senses *sind ja auch um einiges imposanter. Die Wende kam dann ja auch mit* Eco*. Ich traf George Nicholson von Camper & Nicholsons im Rahmen eines Marina-Projekts,*

a few months of bad acts, and mostly out of boredom, I had started drawing yachts and boats.
Then together with the architect John Hix I got employed for a year to develop a housing system for the new town of Milton Keynes. Following this, I got a call from a friend working with Foster to ask me if I could help them with the design of a large glass wall for their first major project, the offices for Willis Faber & Dumas in Ipswich. Over a three-year period I became the "glass man".
Meanwhile my wife and I already had a 8.5-metre Van de Stadt designed sailboat which we had built from a set of mouldings. We had won a few regattas because having built the interior, our boat just happened to be the lightest and I guess we weren't too bad at sailing either (grins). So, we then upgraded to a Contessa 32.
Then 1975, all the architectural projects I was working on stopped – there was a major recession – so we decided to move to the South of France, where my in laws lived. There being no architecture of design work, I started working as a sales agent for Contessa, but I was, and still am, a lousy salesman, so following a suggestion from my wife that we sail round the world and not finding a suitable boat, I decided to design one.

What kind of a boat was that?

A 14 metre sloop I named Prototype*. We built it in a yard in Biot with the aid of a welder and as soon as the hull was finished launched it and went sailing. It had no interior, was very light and went very fast. Somehow, people started talking about it – there weren't really any superyacht media back then – and suddenly I was contracted to design several yachts, a 25-metre ketch and two 26-metre sloops and some smaller ones. At the time, those were big boats. A little while after that, more enquiries came in and I found myself being the designer of the four largest sloops in the world.*

I didn't really know that your career began with sailing yachts.

People tend to forget that today. Eco *or* Senses *are quite a bit more impressive. Things also changed with* Eco*. I met George Nicholson during a marina project that Foster had been invited to do in New York. Foster had said*

►

das er mit Foster umsetzen wollte. Foster zog mich hinzu, weil ich mehr von Booten verstand als er. Als Nicholson mich sah, verstand er zunächst nicht den Zusammenhang zwischen Foster und mir. Als sich die Lage geklärt hatte, entwickelten wir die Marina zusammen mit Peter Rice. Leider verweigerte die Stadt New York die Umsetzung, doch Nicholson hatte für mich noch eine Überraschung parat. Er war, neben dem Marina-Projekt für Azcárraga, zudem der Broker für das neue Schiff von Emilio Azcárraga. Azcárraga war ein mexikanischer TV-Unternehmer, einer der reichsten Unternehmer Süd- und Mittelamerikas. Er hatte die führenden Yachtdesigner der Welt mit Entwürfen für eine Motoryacht zwischen 55 und 65 Metern beauftragt, und Nicholson schob mich da mit hinein.

Ohne Motoryacht-Erfahrung.

Ja, auf den ersten Blick klingt das verrückt. Ich besann mich jedoch auf meine Architektur- und Industriedesign-Vergangenheit, kombinierte dies mit den Segelyacht-Erfahrungen und ging recht unbefangen, aber doch sehr nervös in diesen Pitch. Ich mietete einen Mercedes und wohnte im besten Hotel von Gstaad, wo die Präsentation stattfand. Azcárraga kommentierte zwar, dass mein Design schon sehr gewöhnungsbedürftig war, aber er mochte meine Herangehensweise an solch ein Projekt. Ich bekam den Auftrag. Es war wohl eine Bauchentscheidung des Eigners.

Eco wurde aber über 73 Meter lang und nicht 65 Meter … Wie das so ist mit Yachten, sie werden oft länger als geplant (lacht). Es dauerte jedenfalls eine gewisse Zeit, bis ich den Geschmack von Azcárraga getroffen hatte. Als ich ihm allerdings den Entwurf mit den gebogenen Scheiben zeigte, inspiriert übrigens von Pariser Bussen, war er sofort dabei. Blohm + Voss wurde als Werft auserkoren, weil man dort Erfahrung mit schnellen Schiffen hatte – Eco lief und läuft ja immer noch 35 Knoten. Azcárraga überwies fünf Millionen Dollar auf mein Konto und übertrug mir die Gesamtverantwortung für das Projekt. Ich war Designer, Eignervertreter und Projektmanager in einer Person. Wenn das Geld zur Neige ginge, solle ich wieder melden, sagte Azcárraga.

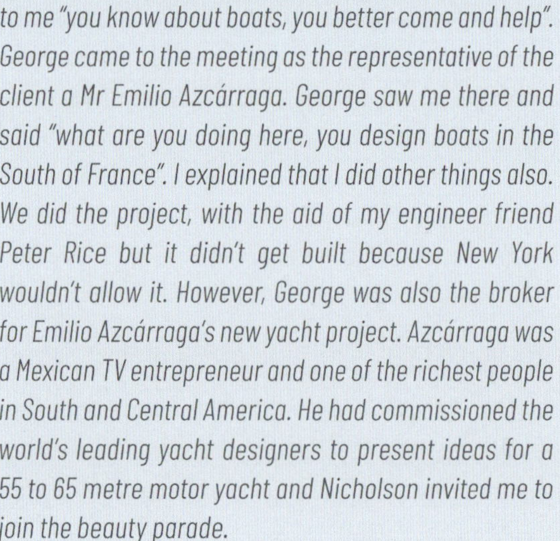

to me "you know about boats, you better come and help". George came to the meeting as the representative of the client a Mr Emilio Azcárraga. George saw me there and said "what are you doing here, you design boats in the South of France". I explained that I did other things also. We did the project, with the aid of my engineer friend Peter Rice but it didn't get built because New York wouldn't allow it. However, George was also the broker for Emilio Azcárraga's new yacht project. Azcárraga was a Mexican TV entrepreneur and one of the richest people in South and Central America. He had commissioned the world's leading yacht designers to present ideas for a 55 to 65 metre motor yacht and Nicholson invited me to join the beauty parade.

Without any experience with motor yachts.

Yes, that sounds crazy when you hear it. But I with my background in architectural and industrial design combined that with my experience working with Foster and my sailing yachts I decided to make a pitch feeling very nervous. I put a lot into it, the presentation was to take place in Gstaad, I hired a Mercedes, stayed in the best hotel and made my presentation to a stony faced Azcárraga, after a while he basically said my design was a heep of junk but he liked my work method so he retained me to develop the project.

But Eco ended up being over 73 metres long instead of 65 … That's the way it can go with yachts, they often end up longer than planned (laughs). It definitely took a while until I managed to find something to Azcárraga's taste. But when I showed him the plans with the curved windows – which, incidentally, were inspired from buses in Paris – he was right on board with it. Blohm + Voss was the chosen shipyard because they had plenty of experience with fast ships – Eco was and still is capable of 35 knots. Azcárraga put five million dollars into my bank account and gave me the overall responsibility for the entire project. I was the designer, owner's rep and project manager all in one. Azcárraga told me that when I needed more money, I was to let him know.

Eine verrückte Zeit.

Wenn man von den fünf Millionen Dollar und den manchmal recht spontanen Entscheidungen hört, natürlich. Das Geschäft war aber direkter, persönlicher. Wir stellten Eco jedenfalls mit einigen Mühen fertig, und Azcárraga liebte diese Yacht fast abgöttisch. Er ließ extra in der Mitte des Atlantiks ein Tankschiff, die Eco Supporter, stationieren, um mit Eco den Ozean zu bezwingen – so riesig war die Reichweite der Yacht nun auch nicht. Bei 30 Knoten schafft sie vielleicht 1.200 Seemeilen nonstop.

Darauf muss man erst einmal kommen.

Es war kreativ, und Geld spielte nicht wirklich eine Rolle. Als Azcárraga schwer krank wurde, zog er sich auf die Yacht zurück und verstarb dort sogar in seiner Kabine. Schön und traurig zugleich. Er war ein toller Kunde und ein Freund zugleich.

Wie ging es für Sie weiter?

Eco war der Start in die Welt der wirklich großen Yachten, wenngleich ich zwischenzeitlich das Architektur- und Ingenieurbüro RFR zusammen mit Peter Rice gründete. Unter anderem war RFR am Louvre und an den Fassaden des National Museum of Science in Paris beteiligt. Peter Rice rief mich dann während meiner Tätigkeit auf Eco schwer krank an und bat mich, bei RFR wieder aktiver zu werden, weil er es nicht mehr schaffte. Ich kam zurück, restrukturierte, und als Peter starb, übergab er mir seine Anteile. Aber dies ist eine andere lange Geschichte. In dieser Phase begann ich zudem, mit Frank Stella zusammenzuarbeiten. Wir sind heute noch eng verbunden.

Stella ist jetzt nicht gerade für seine Yachten bekannt.

Nein, für seine Malerei, für Hard Edge und für seine Skulpturen. Hier kam ich mit meinem Wissensportfolio aus Design, Architektur und Engineering gerade recht. Mit Bootsbauern aus Antibes realisierten wir so einige Stücke. Unter anderem half ich Frank etwa auch, die Skulptur »The Broken Jug« zu realisieren – die misst 11 x 15 Meter und wiegt 45 Tonnen. CMN baute sie in Cherbourg aus Aluminium.

Crazy times.

When you hear about the five million dollars and the at time rather spontaneous decisions, you're right. Business was more direct, more personal, though. Despite a few difficulties along the way, we completed Eco and Azcárraga loved the yacht almost to the point of idolising it. He acquired a refuelling tanker, the Eco Supporter, stationed in the mid Atlantic so that he could cross the ocean at full speed using the gas turbine – the yacht's range at 30 knots plus is around 1,200 miles.

That's an idea you'd first have to come up with.

It was a creative time and unlike now, fees weren't an issue. When Azcárraga fell seriously ill, he sought refuge on the yacht and even died there in his suite. Beautiful yet also sad. He was a fantastic client as well as being a friend.

Where did you go from there?

Eco was my ticket into the world of super yachts, although before it started Peter Rice and I had formed RFR the engineering and architecture practice in Paris where we worked on the National Museum of Science in Paris and the Louvre Pyramid, among other projects. While I was finishing work on Eco, Peter Rice became critically ill and called me up to ask me if I could reassume the role of MD of RFR because he couldn't do it anymore. When he died, he left me half the shares and I had to restructure the company but that is another story. During this time, I also began working with Frank Stella on monumental sculptures and we still have close ties today.

Stella isn't exactly known for his yachts.

No, for his painting, particularly his black paintings, and more recently his sculpture. My knowledge of computer aided design and boat building were a perfect fit for his complex creations, I used artisan boat builders from Antibes to make several pieces. One of the largest projects I helped Frank realise was The Broken Jug sculpture, which was built out of aluminium by CMN in Cherbourg, it is 11 x 15 metres and weighs 45 tonnes.

Sie spielten in dieser Phase in vielen Disziplinen. Wo fühlten Sie sich am wohlsten?
In allen! Ich liebe Kunst, Architektur, Industriedesign und Yachten. Meine Karriere wurde dann Mitte/Ende der 90er-Jahre aber stark von Yachten beeinflusst. Ich arbeitete mit Jack Setton an seiner 59 Meter langen SENSES, die heutzutage immer noch als Synonym für eine Exploreryacht gilt und heute Larry Page von Google gehört. Kurz darauf wurde ich von Larry Ellison engagiert, der ECO gekauft und sie in KATANA umbenannt hatte. Die Yacht bekam ein großes Refit, bei dem wir auch den Hangar für das Wasserflugzeug in einen Basketball-Court umwandelten.

You were active in numerous areas during this phase. Where did you feel most at home?
In all of them! I love art, architecture, industrial design and yachts. But my career was strongly influenced by yachts in the mid to late 1990s. I collaborated with Jack Setton on his 59 metre long SENSES, which is still considered to be the synonym for an explorer yacht and now belongs to Larry Page from Google. Shortly after that Larry Ellison, who had bought ECO and renamed her KATANA, hired me. We completely refitted the yacht, including turning the hangar for the seaplane into a basketball court.

Ellison baute kurz darauf wesentlich größer, nämlich RISING SUN.
Ja, ich half ihm bei den ersten Ideen, dann entschied er sich, mit Jon Bannenberg zu arbeiten. So ist das bei uns in der Szene. Mal ist man dabei, mal nicht. Als ich das erste Design für Ellison für einen türkischen Eigner weiter ausarbeitete, hatte ich ebenfalls Pech. Ich wunderte mich schon, warum ich nichts mehr von ihm gehört hatte, als ich in der Zeitung davon las, dass Nokia und Ericsson ihn auf sechs Milliarden Dollar verklagten. Da gab es keinen Spielraum für eine 126-Meter-Yacht. Die nächsten Projekte, für die ich angefragt wurde, waren indes nicht weniger spektakulär.

Ellison later built a much larger yacht, the RISING SUN.
That's right. I helped him with the initial ideas and then he decided to work with Jon Bannenberg. That's the way it is in our business. Sometimes you're in, sometimes you're not. When I further developed the designs I had made for Ellison for a Turkish owner, I was also unlucky. I was wondering why I hadn't heard back from him when I read on the front page of the Financial Times that Nokia and Ericsson were suing him for a great deal of money. That didn't leave much for a 126-metre yacht. The next projects I was asked to do in the meantime weren't any less spectacular.

Und zwar?
Irgendwann, als ich mein Studio in London hatte, sagte meine Assistentin, dass ein gewisser Mister Jobs in der Leitung wäre. Ich fragte: »Jobs von Apple?« Sie zuckte mit den Schultern, und als sie ihn durchgestellt hatte, fragte ich: »Sind Sie es, Steve?« Er sagte: »Ja. Wann haben Sie Zeit, nach Kalifornien zu komme? Ich kann nicht reisen, bin etwas krank.« Ich traf ihn daraufhin einige Male, es war eine wundervolle Erfahrung. Dass er letztlich mit Philippe Starck die VENUS realisierte, war ok. Starck und ich hatten zusammen ja eine andere schöne Baustelle.

What were they?
At one point, when I had my office in London my PA told me a certain Mr Jobs was on the phone. I asked: "Jobs from Apple?" She didn't know and put him through, I said: "Is it you Steve?" He replied, "Yes, I want to find out about yachts and have been ill so I can't travel, would you have time to come to California?" I went to visit him in his home several times, it was a unique experience. The fact that he ended up building the VENUS with Philippe Starck was fine. Starck and I were already working on another project together.

Die Motoryacht A.

Exakt. In der Entwicklungs- und Bauphase hieß sie SF99, wobei die Buchstaben für Starck und Francis stehen. Starck kam mit diesem wirklich außergewöhnlichen Design, hatte aber keine Ahnung, ob so etwas überhaupt schwimmen würde. Ich bekam quasi das technische Design übertragen und investierte sehr viel Zeit und Geld in Research & Development. Die Bedenken bezüglich des negativen Stevens der Yacht war zunächst groß, kehrten sich dann aber ins Gegenteil um. A funktioniert allein durch ihren Bug und ihre extrem lange Wasserlinie fantastisch. Die Zusammenarbeit mit Starck war herrlich und interessant.

Sind Sie eigentlich, rückblickend, ein wenig enttäuscht oder eifersüchtig, dass einige Ihrer Schüler oder Mitbewerber wesentlich mehr Yachten realisiert haben als Sie?

Harte Frage, aber berechtigt. Es dauerte eine gewisse Zeit, bis ich mich damit arrangiert hatte. Allerdings habe ich inzwischen eine sehr große Zufriedenheit erlangt, weil einige meiner Ablieferungen fast prägend sind. Nehmen Sie Eco, Senses oder auch A, oder ungebaut Sultan und Crystal Ball. Und nur aus Yachten bestand mein Arbeitsleben auch nicht.

Ich könnte jetzt noch Golden Odyssey erwähnen oder den Tender, den Sie mit Mercedes-Benz launchten.

Natürlich. Allerdings ist Golden Odyssey ein Projekt, über das nicht viel gesprochen werden darf; und über den Daycruiser von Silver Arrows Marine, also das Mercedes-Boot, wurde doch schon sehr viel geschrieben.

Noch ein Rat zum Schluss? Eine Lebensweisheit von Martin Francis?

Wer wie ich mitunter von Selbstzweifeln geplagt wird, muss sich gut in seiner Haut fühlen, um erfolgreich zu sein (was ich nicht wirklich bin). Manche Menschen verdienen viel Geld, andere Menschen produzieren Dinge. Wer Dinge produziert, verdient meist nicht viel Geld. Und ich produziere lieber Dinge.

Dieses Interview erschien in veränderter Form bereits in BOOTE EXCLUSIV.

Motoryacht A.

Exactly. During development and construction, she was called SF99; the letters stood for Starck and Francis. Starck came up with this really unusual design, but he didn't have a clue whether it would even float. So I was essentially given the responsibility for the technical design and preliminary naval architecture. The client invested a great deal of time and money in research and development, because there were initially a lot of doubts about the yacht's wave-piercing hull but in the end that changed completely after we had completed the tank tests. A runs fantastically thanks to her extremely fine bow entry; working with Starck was both enjoyable and instructive.

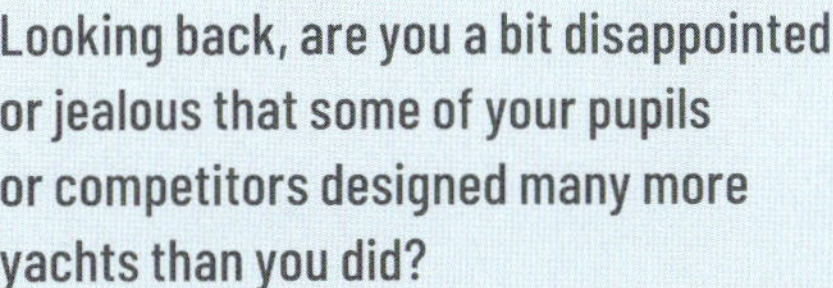

Looking back, are you a bit disappointed or jealous that some of your pupils or competitors designed many more yachts than you did?

That's a difficult question but justified. It took a while until I came to terms with that. But I'm now really content because some of my projects have been influential. For example, Eco, Senses, A or unbuilt, Sultan and Crystal Ball. After all my career doesn't just only consist of yachts.

I could mention Golden Odyssey as well or the project with Mercedes-Benz Style.

Of course. However, Golden Odyssey is a project that I'm not allowed to say much about; and a lot has already been written about the Gran Turismo from Silver Arrows Marine, the Mercedes boat.

Any advice to finish off with? Some worldly wisdom from Martin Francis?

As one who is often filled with self doubt, I think that you have to feel good in your skin to be really successful (which I am not). After all, some people make money, some people make things, people who make things rarely make money, I make things.

This interview was published in slightly different version in BOOTE EXCLUSIV.

Inspiration
Inspiration

Designer wie Martin Francis sind ständig auf der Suche nach inspirierenden Eindrücken aus anderen »Disziplinen« – Autos, Flugzeuge, Schiffe, Gebäude oder schlicht die Natur gehören dazu.

Für Martin Francis war unter anderem der Crystal Palace ein wichtiger Denkanstoß. Das Gebäude, von dem Gärtner Joseph Paxton 1851 für die damalige Weltausstellung in London gestaltet, besitzt eine immense, 84.000 Quadratmeter messende Glasstruktur – die größte, die bis dato realisiert wurde. Nicht minder begeistert ist Francis von seiner Sammlung sogenannter Forcolas, den Gabeln für die Riemen der venezianischen Gondeln und natürlich von allen Produkten, deren Design streng ihrer Funktion folgt. Solche Objekte faszinieren ihn und lassen ihn querdenken. Pariser Busse führten etwa zum Design der Megayacht-Ikone Eco.

Designers like Martin Francis are constanly looking to other disciplines and historical references for inspiration, be it in nature, aeroplanes, automobiles, ships or buildings.

For Martin Francis, the Crystal Palace built in London for the great exhibition of 1851 is one of the most important references. Designed by the gardener Joseph Paxton, and built in less than a year, the building was covered with 84,000 square metres of a superbly detailed glass envelope. He is no less enthusiastic about his collection of forcole the rowlocks used in traditional Venetian boats, of course all products whose form follows their function. Such references often make him think outside the box. Paris buses, for example, led to the window design of the iconic mega yacht Eco.

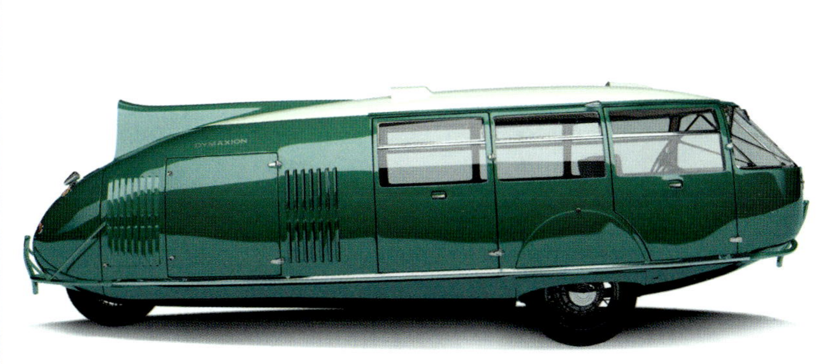

CITY OF ST. PETERSBURG

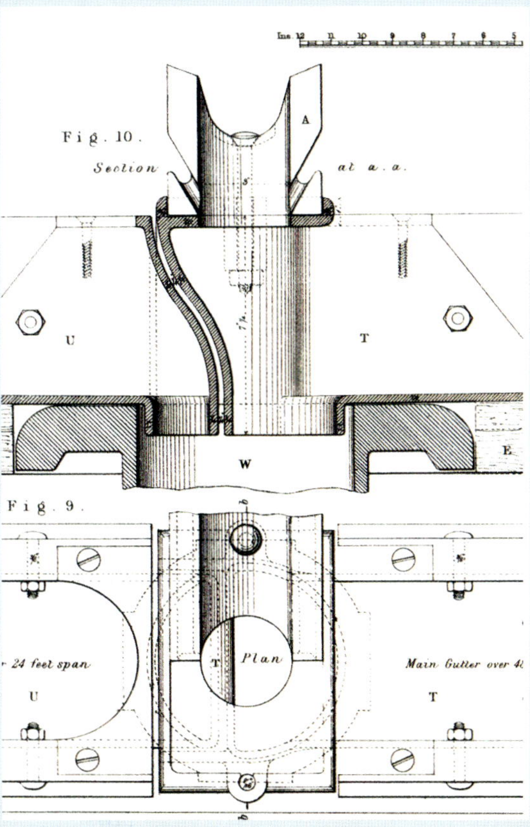
Fig. 10.
Section
at a. a.
A
U
T
W
E
Fig. 9.
Plan
U
T

»Ein Boot ist ein in sich selbst abgeschlossenes Objekt. Man kann im Prinzip bauen, was man will, wenn man den richtigen Kunden dafür hat.«

“A boat is a self-contained object. In principle you can build what you want, as long as you have a willing client.”

Yachten
Yachts

46–57

26–37

42–43

38 – 41

66 – 71

58 – 65

72 – 77

ECO | ENIGMA | KATANA | ZEUS

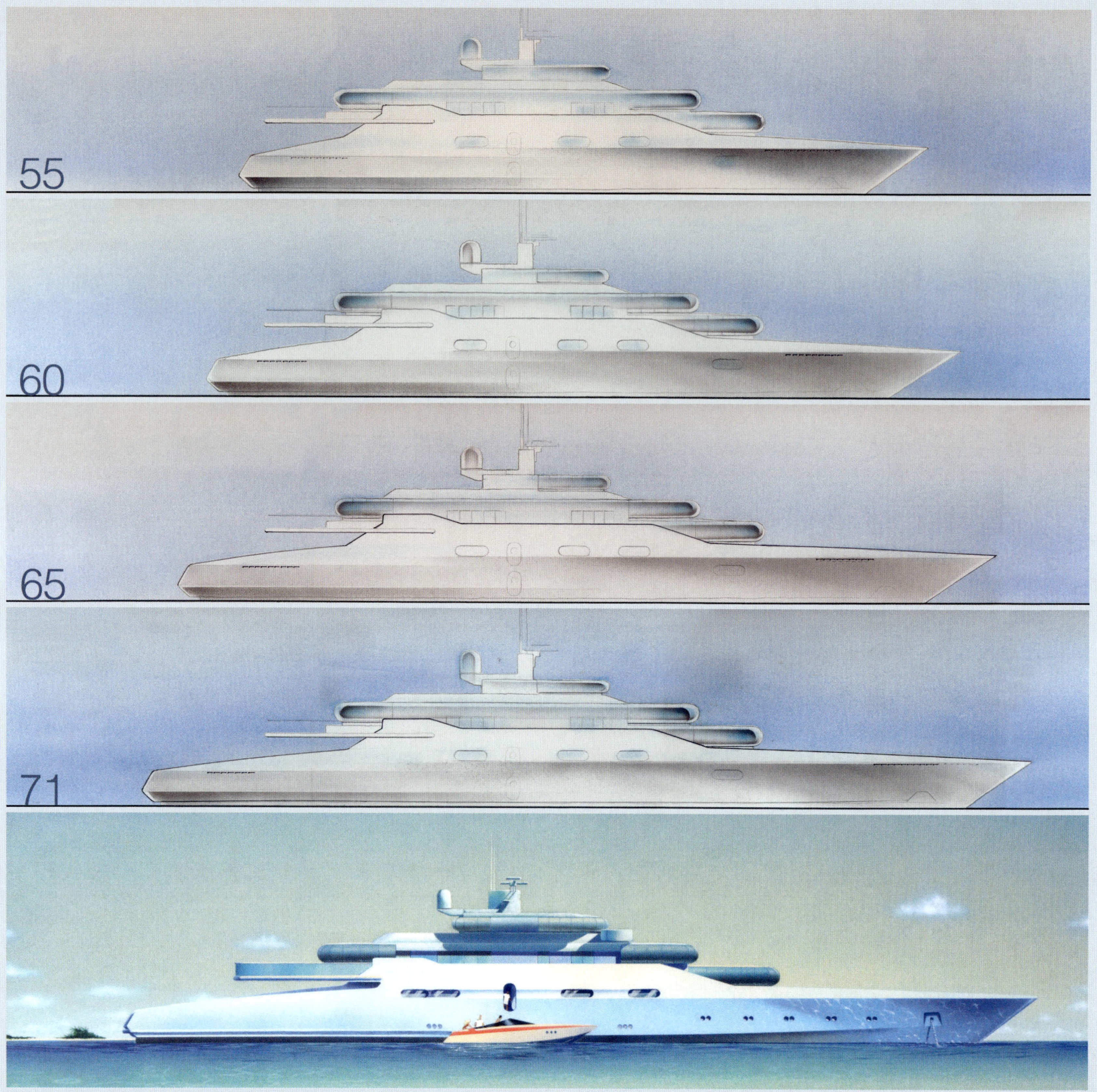
55
60
65
71

GEBURT EINER IKONE

Das Briefing sah eigentlich eine Yacht zwischen 55 und 65 Metern vor. Je länger der Eco-Auftraggeber Emilio Azcárraga und Martin Francis jedoch dessen Entwürfe diskutierten, um so länger wurde schließlich die Yacht – 74,50 Meter bei nur 11,20 Meter Breite. Den mexikanischen Milliardär überzeugte dabei nicht nur Francis' Gestaltung, sondern auch seine ganzheitliche Denkweise. Für den Bau der 35 Knoten schnellen Yacht bei Blohm + Voss übertrug er dem Designer das komplette Projektmanagement samt Budgetverantwortung. Aussage Azcárraga: »Wenn die fünf Millionen Dollar verbraucht sind, rufen Sie einfach an.«

BIRTH OF AN ICON

The original brief envisaged a yacht between 55 and 65 metres. However, the longer the Eco client Emilio Azcárraga and Martin Francis discussed the design, the longer the yacht became – 74.50 metres with a width of only 11.20 metres. The Mexican billionaire was convinced not only by Francis' design, but also by his holistic way of thinking. For the construction of the 35 knots fast yacht at Blohm + Voss he entrusted the designer with the complete project management including budget responsibility. To start the contract, Azcárraga gave five million dollars to Francis and said: "When you need more, just give me a call."

INSPIRATION AUS DEM NAHVERKEHR

Ihre gebogenen Scheiben sind das Kennzeichen von Eco, die durch mehrere Eignerwechsel erst KATANA und dann ENIGMA hieß und jetzt auf den Namen ZEUS hört. Der Look erstreckt sich über drei Decks, manche der Gläser fungieren auch als Türen. Die Inspiration nahm Francis von Pariser Bussen, deren Scheiben zur Reflexionsminderung eingesetzt wurden; als Hersteller wurde Flachglas engagiert. Die 19 Millimeter dicken und drei Quadratmeter großen Komponenten waren, so Francis, »eine Entwicklung ins Blaue«. Trotz der Auflage der Klassifikationsgesellschaft, Rahmenverstärkungen für die Gläser mitzuführen, kamen diese nie zum Einsatz.

INSPIRATION FROM PUBLIC TRANSPORT

The curved glass panes are the hallmark of Eco, which was then called KATANA and subsequently ENIGMA and after several changes of ownership, is now named ZEUS. The glazing extend over three decks and some of the panels also serve as doors. Francis took his inspiration from Parisian buses where they where used in the 1970s to reduce internal reflections, Flachglas in Germany was the selected supplier. According to Francis, the 19 millimetre thick, three square metre 750 millimetre radius curved and toughend panels were breaking new ground. Despite the classification society's requirement to extra mullions for the wheelhouse glazing, they were never used.

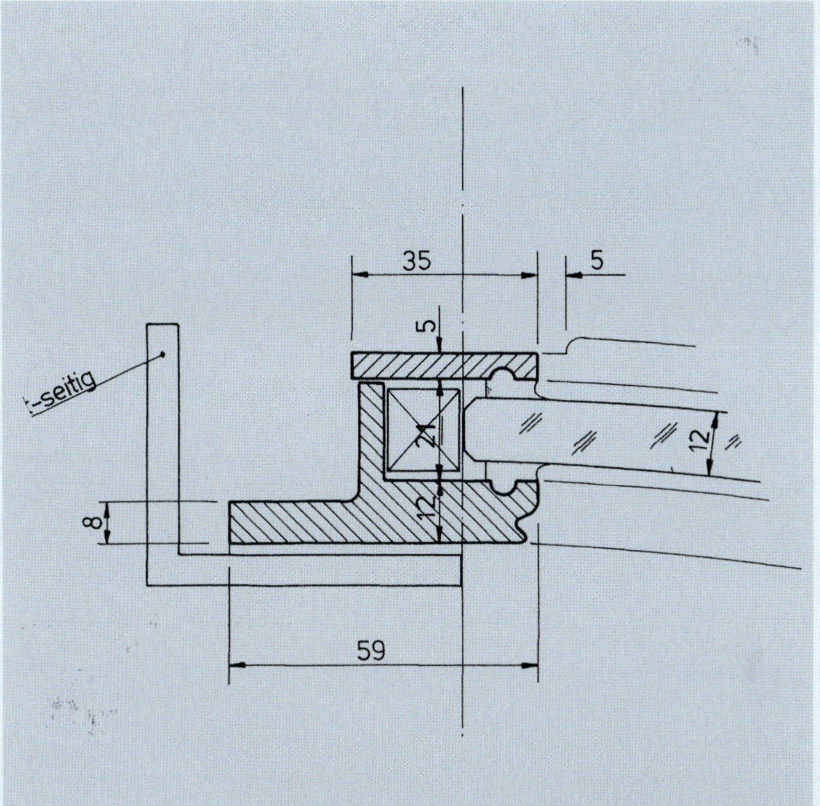
35
5
5
-seitig
24
12
12
8
59

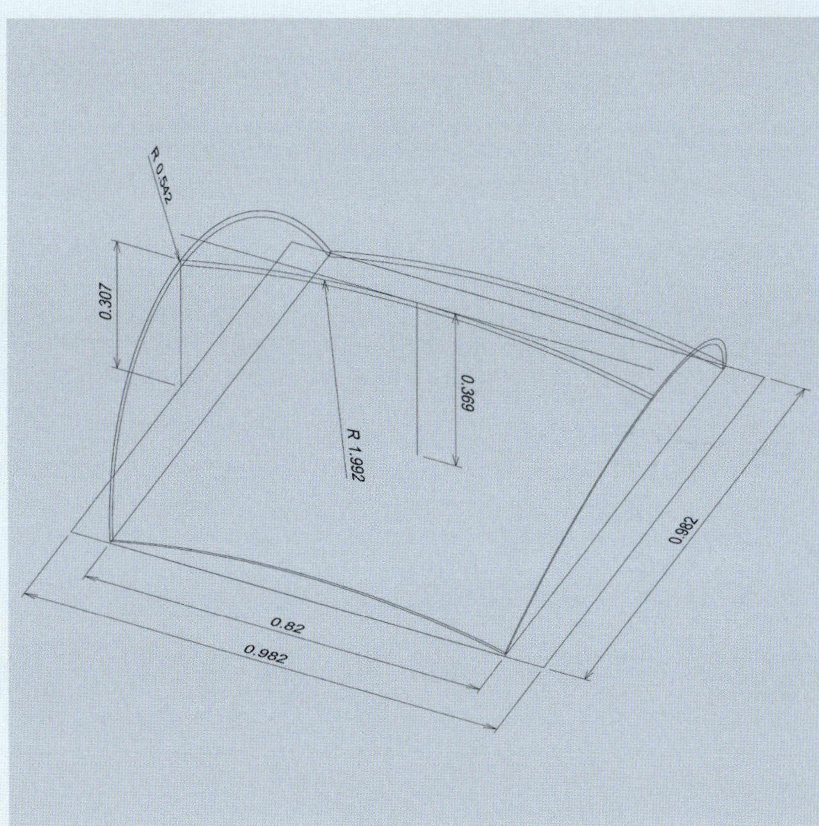
R 0.542
0.307
0.369
R 1.992
0.982
0.82
0.982

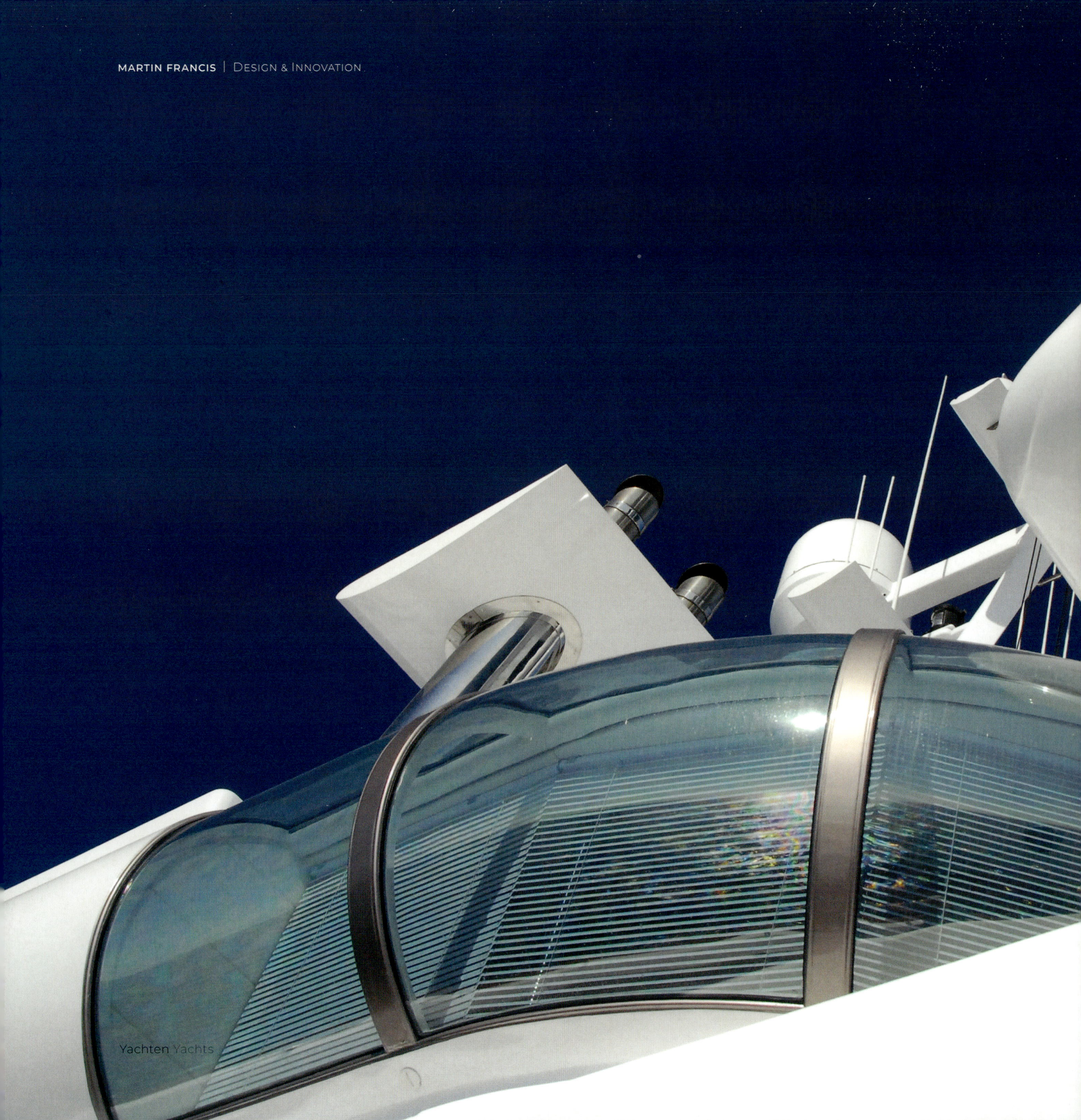

ZEITLOSER LOOK AUS DEN 1990ER-JAHREN

Das Interieur von Eco schuf Francis zusammen mit seinem Kollegen François Zuretti, für den dies das erste Yachtprojekt war. Aufgrund der flachen Silhouette und der geringen Breite der Yacht sind die Räumlichkeiten etwas gemütlicher als auf den heutigen, eher voluminösen 70-Meter-Yachten. Die Auswahl der Hölzer und Stoffe wirkt damals wie heute modern; der Blick aus den konvexen Scheiben ist spektakulär. In der Eignerkabine, die sich auf dem Oberdeck befindet, zog sich Emilio Azcárraga in den letzten Monaten seines Lebens zurück und verstarb an Bord.

TIMELESS LOOK FROM THE 1990S

The interior of Eco was created by Francis together with François Zuretti for whom it was the fist yacht. Due to the flat silhouette and the small width of the yacht, the rooms are somewhat more cozy than on today's rather voluminous 70-metre yachts. The choice of woods and fabrics was as modern then as it is today; the view from the convex windows is spectacular. In the owner's cabin on the upper deck, Emilio Azcárraga spent the last months of his life and died on board. It was his favourite place to live.

ENIGMA

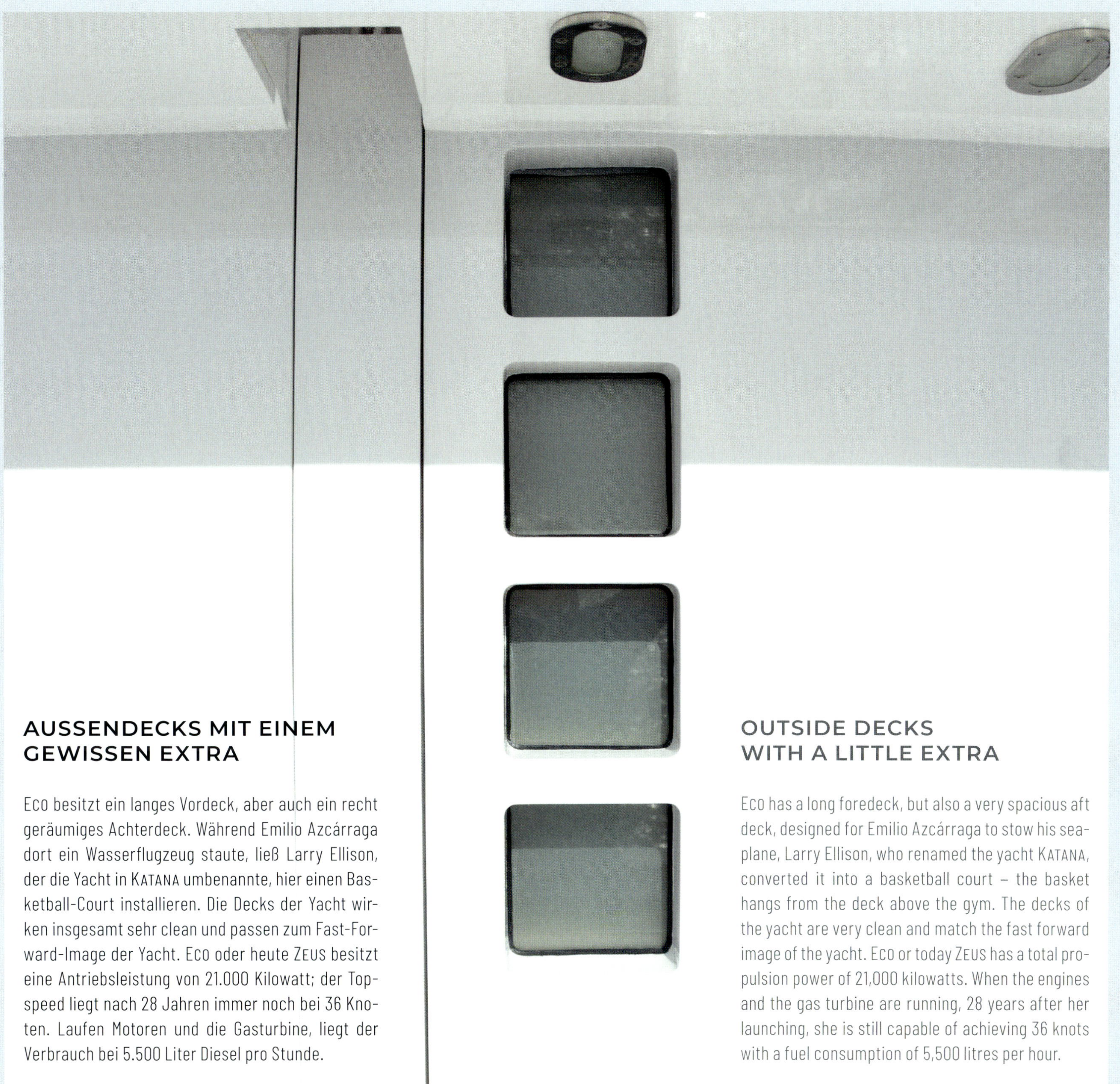

AUSSENDECKS MIT EINEM GEWISSEN EXTRA

Eco besitzt ein langes Vordeck, aber auch ein recht geräumiges Achterdeck. Während Emilio Azcárraga dort ein Wasserflugzeug staute, ließ Larry Ellison, der die Yacht in Katana umbenannte, hier einen Basketball-Court installieren. Die Decks der Yacht wirken insgesamt sehr clean und passen zum Fast-Forward-Image der Yacht. Eco oder heute Zeus besitzt eine Antriebsleistung von 21.000 Kilowatt; der Topspeed liegt nach 28 Jahren immer noch bei 36 Knoten. Laufen Motoren und die Gasturbine, liegt der Verbrauch bei 5.500 Liter Diesel pro Stunde.

OUTSIDE DECKS WITH A LITTLE EXTRA

Eco has a long foredeck, but also a very spacious aft deck, designed for Emilio Azcárraga to stow his seaplane, Larry Ellison, who renamed the yacht Katana, converted it into a basketball court – the basket hangs from the deck above the gym. The decks of the yacht are very clean and match the fast forward image of the yacht. Eco or today Zeus has a total propulsion power of 21,000 kilowatts. When the engines and the gas turbine are running, 28 years after her launching, she is still capable of achieving 36 knots with a fuel consumption of 5,500 litres per hour.

GOLDEN SHADOW

DAS ERSTE SHADOWBOOT DER WELT

Mitte der 1990er-Jahre musste Francis oft nach San Diego fliegen. Der Eigner der ersten GOLDEN ODYSSEY ließ dort die GOLDEN SHADOW bauen. Die 65 Meter lange Yacht war als reines Support Vessel konzipiert, um alle möglichen Beiboote sowie gar ein Wasserflugzeug für die GOLDEN ODYSSEY zu transportieren. Damit war der Auftraggeber seiner Zeit weit voraus: GOLDEN SHADOW war mit Abstand die erste Yacht, die lediglich diesen Zweck erfüllte. Heute bieten mehrere Werften diese sogenannten Support Vessel an. Wenn sie für ihr Mutterschiff die Tender, den Helikopter und vielleicht auch einen Teil der Besatzung transportieren, bleibt auf der eigentlichen Yacht mehr Volumen für den Eigner und seine Gäste.

THE WORLD'S FIRST SHADOW BOAT

In the mid-1990s Francis often had to fly to San Diego. The owner of the first GOLDEN ODYSSEY, delivered by Blohm + Voss in 1990, had the GOLDEN SHADOW built there. The 65 metres long yacht was designed as a pure support vessel to transport a variety of tenders and even a Cessna Caravan float plane for the GOLDEN ODYSSEY. Thus the client was way ahead of his time: GOLDEN SHADOW was the first yacht that only served this purpose. Today, several shipyards offer these so-called shadow boats or support vessels. They transport the tenders, the helicopter and perhaps also a part of the crew for their mother ship, so more volume remains on the mother ship for the owner and his guests.

WASSERFLUGZEUG IN DER FLOTTE

Der ungewöhnlichste Tender, den Golden Shadow in ihrem Heck transportierte, war – neben den insgesamt sechs Beibooten – das Wasserflugzeug »Golden Eye«. Es komplettierte die sogenannte Goldene Flotte, zu der neben der Golden Shadow und der Golden Odyssey auch noch die Golden Osprey gehörte, ein 30 Meter langes Spezialboot zum Hochseeangeln.
Auf Golden Shadow ließ der Eigner ein Labor für maritime Forschungen installieren, das zahlreiche Universitäten und Forschungseinrichtungen nutzten. Heute steht die Yacht mit einer Reichweite von 13.000 Seemeilen solventen Charterurlaubern zur Verfügung.

SEAPLANE IN THE FLEET

The most unusual tender, which Golden Shadow transported in its stern, was – besides the six boats on deck – certainly the seaplane "Golden Eye". It compleats the so-called Golden Fleet, to which the Golden Shadow and the Golden Odyssey belong as well as the Golden Osprey, a 30 metres boat long for deep sea fishing also remodeled by Francis.
On Golden Shadow the owner had a laboratory for maritime studies installed, which was used by numerous universities and research institutions, the yacht with a range of 13,000 nautical miles is available to wealthy charterguests.

GOLDEN ODYSSEY

EINE NACHFOLGERIN NACH 25 JAHREN

GOLDEN ODYSSEY ist bislang der einzige Francis-Entwurf, der bei Lürssen, der vielleicht produktivsten Werft für 100-Meter-plus-Yachten, gebaut wurde. Die 124 Meter lange und 20 Meter breite Yacht entstand als Projekt TATIANA für einen Bestandskunden von Francis, der seine erste GOLDEN ODYSSEY 1993 in San Diego umbauen ließ. »Und bereits damals sprachen wir über eine Nachfolgerin«, sagt Francis, der die neue GOLDEN ODYSSEY gern als »längste Entwicklung meiner Karriere« bezeichnet. Die 2017 gewasserte Yacht nimmt in ihren 16 Kabinen 32 Gäste auf.

A SUCCESSOR AFTER 25 YEARS

GOLDEN ODYSSEY is so far the only Francis design built by Lürssen, perhaps the most productive shipyard for 100 metres plus yachts. The 124 metres long and 20 metres wide yacht was built as a project TATIANA for an existing client of Francis, who had converted the first GOLDEN ODYSSEY in San Diego in 1993. "Even then we talked about a successor," says Francis, who likes to call the new GOLDEN ODYSSEY the "longest development of my career". Launched in 2017 she can accommodate 32 guests in 16 cabins at a length of 124 metres.

»Technik ist eine Form der Kunst. Und ich wage von mir zu behaupten, dass ich ein Designer bin und kein Stylist.«

"Technology
is a form of art,
I like to think
I'm a designer
not a stylist."

MARTIN FRANCIS | DESIGN & INNOVATION

MOTORYACHT A

Yachten Yachts

EIN GANZES JAHR FÜR DIE ENTWICKLUNG

Der Rumpf der 119 Meter langen A sieht radikal und etwas unwirklich aus, bewährt sich in der Praxis jedoch außerordentlich gut. Zum einen ist er an die Zumwalt-Klasse amerikanischer Zerstörer angelehnt, zum anderen testeten Francis und sein Team ihn ein ganzes Jahr lang. Der sehr lange Steven erzeugt beispielsweise um so weniger Wellen, je schneller die Yacht fährt. Bei Probefahrten mit einem motorisierten Modell auf dem südenglischen Solent überstand die A sogar Wellen, die – skaliert betrachtet – um die zehn Meter hoch waren. Um die Glasflächen des Aufbaus zu schützen, forderte die Klassifikationsgesellschaft Lloyds dennoch einen versenkbaren (und noch nicht genutzten) Wellenbrecher auf dem Vordeck.

A WHOLE YEAR FOR DEVELOPMENT

The hull of the 119 metre long A looks radical and somewhat unreal, but proves itself in practice extraordinarily well. On the one hand it is based on the Zumwalt class of American destroyers, on the other hand Francis and his team tested it for a whole year. The very long stem, for example, generates fewer waves the faster the yacht is. During test drives with a motorised and approximately three metre long model on the southern English Solent, A even survived waves that, when scaled, were about ten metres high. In order to protect the glass surfaces of the superstructure, the classification society Lloyds nevertheless demanded a retractable (and not yet used) breakwater on the foredeck.

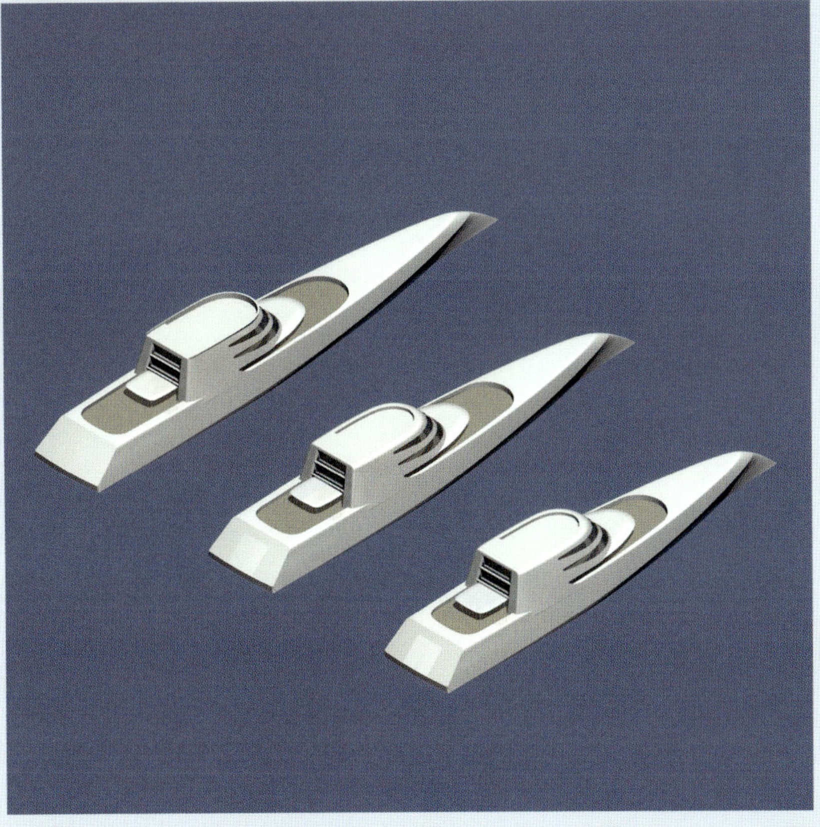

AUF DEM VORDECK LANDET DER HELIKOPTER

Die vielleicht ungewöhnlichste Yacht, die je gebaut wurde, schneidet förmlich durch das Wasser. Für den Auftrieb des schmalen Vorschiffs sorgt ein sogenannter Wulstbug, eine Stahlnase unter Wasser. Den Platz über der verdeckten Ankertechnik kann ein Helikopter zur Landung nutzen. Angetrieben wird A, an der die Kieler Werft HDW vier Jahre lang baute, von zwei MAN-Motoren des Typs RK280. Die insgesamt 18.000 Kilowatt sorgen für einen Topspeed von 23 Knoten. Bei 19 Knoten kann die Yacht 6.500 Seemeilen – also 12.000 Kilometer – nonstop mit einer Tankfüllung (757.000 l) fahren.

THE HELICOPTER LANDS ON THE FOREDECK

Perhaps the most unusual yacht ever built cuts through the water. The buoyancy of the narrow bow is provided by a so-called bulbous bow, a steel nose under water. A helicopter can use the space above the concealed anchor technology for landing. A, on which the Kiel based shipyard HDW built for four years, is powered by two MAN RK280 engines. The total of 18,000 kilowatts ensures a top speed of 23 knots. At 19 knots, the yacht can sail 6,500 nautical miles – i. e. 12,000 kilometres – non-stop with one tank of fuel (757,000 litres).

IN DEN AUFBAUTEN WOHNT DAS EIGNERPAAR

Die Silhouette der A gleicht keiner anderen Yacht. Philippe Starck, Chef Concept-Designer von A, mit dem Francis das Projekt unter dem Code SF99 entwickelte, verzichtete bei seiner Gestaltung auf allzu viel umbauten Raum und verordnete A einen Aufbau, der an den verglasten Turm eines U-Bootes erinnern könnte. Darin befinden sich die Eignersuite samt Jacuzzi und einem Bett auf drehbarem Sockel, der Salon – ebenfalls mit Jacuzzi – und der Arbeitsplatz des Kapitäns. Die Gäste wohnen in sechs Kabinen auf dem Unterdeck; ihre Lage erkennt man an den ovalen Fenstern im Rumpf (siehe vorige Doppelseite).

THE OWNER COUPLE LIVES IN THE TOWER

The silhouette of A does not resemble any other yacht. Philippe Starck, Chief Concept Designer of A, with whom Francis developed the project under the code SF99, dispensed with too much enclosed space in his design and ordered her to have a superstructure reminiscent of the glazed tower of a submarine. It contains the owner's suite with a Jacuzzi and a bed on a revolving pedestal, the saloon – also with Jacuzzi – and the captain's workstation. The guests sleep in six cabins on the lower deck; their position can be recognised by the oval windows in the hull (see previous double page).

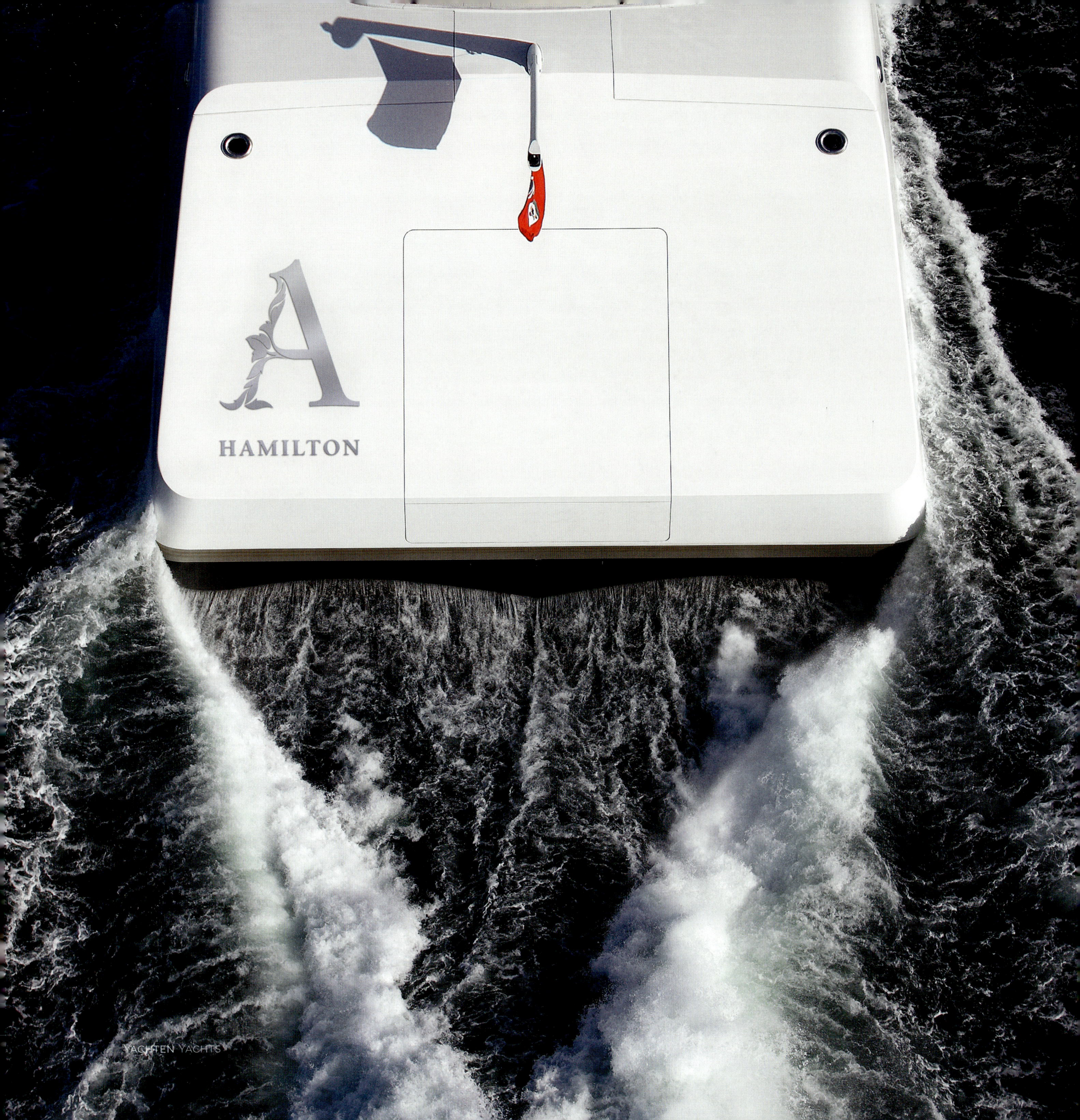
A
HAMILTON

BLICKFANG AUF ALLEN WELTMEEREN

Neben der Heckklappe besitzt A insgesamt sieben Luken auf jeder Seite. Diese Anzahl bedurfte starker konstruktiver Überlegungen, um die Stabilität des Stahlrumpfes nicht zu beeinträchtigen. Die größten Klappen sind 14 Meter breit und 4,50 Meter hoch. Dahinter befinden sich die Vaudrey-Miller-Tender, selbstverständlich Einzelanfertigungen, gebaut in Neuseeland. Um die Formen des Entwurfs nicht zu stören, wurden möglichst viel Technik und Zubehör – also Anker, Antennen, Kräne – hinter Klappen versteckt. Wo A auch anlegt, sie sorgt für Aufsehen.

EYE-CATCHER ON ALL OCEANS

In addition to the aft hatch, A has a total of seven hatches on each side. This number required a lot of constructive considerations in order not to impair the stability of the steel hull. The largest flaps are 14 metres wide and 4.50 metres high. Behind them are the Vaudrey Miller tenders stowed, of course custom-made, built in New Zealand. In order not to disturb the forms of the design, as much technology and accessories as possible – i. e. anchors, antennas, cranes – were hidden behind flaps. Wherever A docks, she is a head turner.

SENSES

ABFAHRBEREIT NACH NUR 18 MONATEN BAUZEIT

Die 59,20 Meter lange SENSES gilt als Vorbild in der Gattung der Exploreryachten, also Yachten, die durch ihre Bauweise und ihre Ausstattung nahezu jedes Revier der Weltmeere befahren können. Auftraggeber für die Yacht war der Franzose Jack Setton, ein erfahrener Eigner, der zuvor die 77,70 Meter lange SIMSON S besessen hatte. Gebaut wurde SENSES in nur 18 Monaten bei Stahlbau Nord (Bremerhaven) und bei der Schweers-Werft (Elsfleth). Während Martin Francis für das Exterior und die Konstruktion verantwortlich war, gestaltete Philippe Starck die Inneneinrichtung.

READY TO GO AFTER ONLY 18 MONTHS

The 59.20 metre long SENSES is regarded as a model in the class of explorer yachts, i. e. yachts which can sail almost any area of the world's oceans due to their construction and equipment. The client for the yacht was the Frenchman Jack Setton, an experienced owner who had previously owned the 77.70 metre long SIMSON S. SENSES was built in just 18 months by Stahlbau Nord (Bremerhaven) and Schweers-Werft (Elsfleth). While Martin Francis was responsible for the exterior and construction, Philippe Starck designed the interior.

MIT EINER ARMADA VON TENDERN BESTÜCKT

Wo die Buchten zu flach oder die Häfen zu klein für SENSES waren, konnte Jack Setton mit Beibooten weiterfahren. Im Heck der Yacht transportierte er ein 11,60 Meter langes Festrumpf-Schlauchboot mit 60 Knoten Topspeed, ein Rettungsboot, ein sieben Meter langes Halmatic, Jetskis, einen Hobie-Katamaran, einen Herreshoff-Segler und einen Helikopter. Zudem lagerte reichlich Tauch-Equipment an Bord. Highlight der Flotte war indes eine 12,80-Meter-Motoryacht mit 14,5 Tonnen Gewicht, die per Fernbedienung aus dem Heck heraus gewassert werden konnte.

EQUIPPED WITH AN ARMADA OF TENDERS

Where the bays were too shallow or the harbours too small for SENSES, Jack Setton could continue with numerous dinghies. In the stern of the yacht he transported, among other things, an 11.60 metres long inflatable rigid hull with 60 knots top speed, a lifeboat, a seven metres long Halmatic, jet skis, a Hobie catamaran, a Herreshoff sailboat and a helicopter. There was also plenty of diving equipment on board. The highlight of the fleet was a 12.80 metre motoryacht weighing 14.5 tons, which could be pushed out of the stern by remote control via a ramp.

TK IRS

MARTIN FRANCIS | DESIGN & INNOVATION

SILVER ARROW 460

Yachten Yachts

EIN DAYBOAT MIT DEM DAIMLER-STERN

Der Mercedes-Stern steht für »zu Lande, zu Wasser und in der Luft«. Als Gottlieb Daimler vor 125 Jahren das erste Motorboot der Welt baute, die Marke dann mit Autos weltbekannt wurde und zusammen mit Airbus einen Helikopter umsetzte, fehlte noch ein Boot zur Portfolio-Ergänzung. Mercedes-Benz Style ergriff die Initiative und stellte ein Team zusammen, das zunächst aus den Projektmanagern von Silver Arrows Marine, Martin Francis und Tommaso Spadolini bestand. Zusammen entwickelte man einen 14 Meter langen Prototyp, um die Machbarkeit zu testen. Im Jahr 2016 wurde der erste »Silberpfeil« gewassert.

A DAYBOAT WITH THE DAIMLER STAR

The three-pointed Mercedes star stands for "on land, at sea and in the air". After Gottlieb Daimler built the world's first motorboat more than 125 years ago, the brand became world-famous for its cars and, together with Airbus, built a helicopter, but a boat to complement the portfolio was still missing in this millennium. Mercedes-Benz Style took the initiative and put together a team consisting of the project managers of Silver Arrows Marine, Martin Francis and Tommaso Spadolini. Together they developed a 14 metres long prototype to test the feasibility. In 2016, the first "Silver Arrow" was launched.

DIE DESIGNPHILOSOPHIE? »IT'S HOT AND COOL«

Im Gegensatz zum Automobil-Design, wo sich das fertige Produkt sehr von ersten Entwürfen unterscheiden kann, änderte sich an der SILVER ARROW 460 über den gesamten Entwicklungsprozess relativ wenig. Das Boot mit markanter »Dropping Line« wird komplett aus Carbon gebaut – inzwischen bei Baltic Yachts – verdrängt 13 Tonnen und schafft mit zwei je 350 kW starken Motoren eine Höchstgeschwindigkeit von 40 Knoten. Gorden Wagener, Designchef der Daimler AG, sagte über Martin Francis: »Er gilt zu Recht als einer der Besten seiner Branche und ist ein supernetter Typ.«

THE DESIGN PHILOSOPHY? "IT'S HOT AND COOL"

In contrast to the automobile design, where the finished product can be very different from the first designs, the SILVER ARROW 460 changed relatively little during the entire development process. The silver-painted boat with a striking "Dropping Line" is built entirely of carbon – now at Baltic Yachts in Finland – and displaces 13 tons. With two 350 kW engines each, it achieves a top speed of 40 knots. Gorden Wagener, head of design at Daimler AG, said of the cooperation with Martin Francis: "He is rightly regarded as one of the best in his industry and is a super nice guy."

Mercedes-Benz

Yachten Yachts

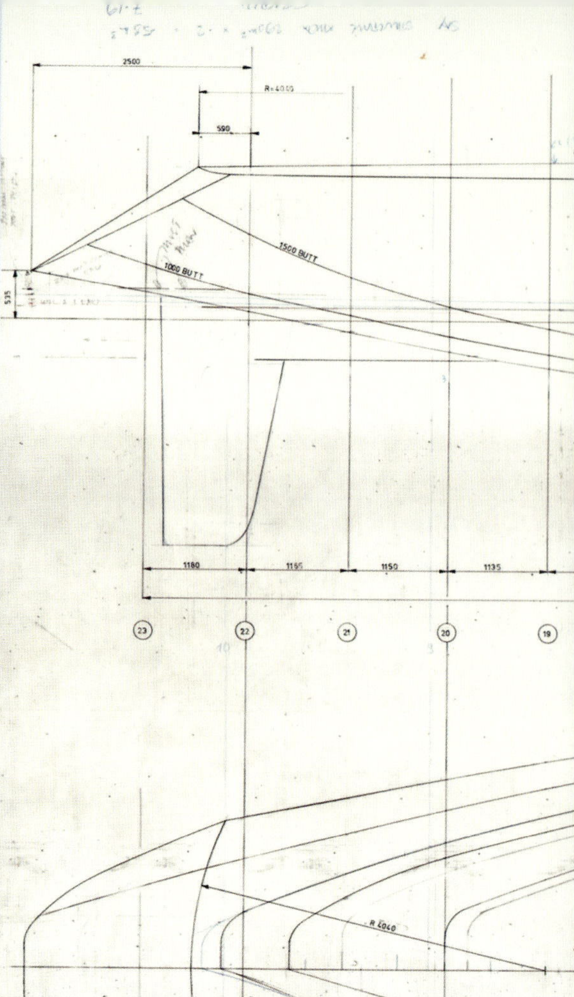

DIABLESSE

DAS DAMALS GRÖSSTE EINZELRIGG DER WELT

Durch die Wasserung der 28,20 Meter langen Diablesse im Jahr 1981 zählte Martin Francis für eine kurze Zeit als Designer der weltgrößten Slups, also Segelyachten mit nur einem Mast – neben Diablesse entstanden auch noch Must, Chrismi II und La Concorde. Die Yachten wurde in Südfrankreich bei Trehard und Chantier Navals de l'Esterel gebaut. Diablesse umrundete nach zahlreichen gewonnenen Regatten bereits dreimal die Welt. Ihr Liftkiel lässt sie auch flache und entlegene Reviere erreichen; bis zu acht Gäste können in drei Kabinen an Bord schlafen. Der Auftrag für Diablesse war ein Resultat von Prototype, einer 14 Meter langen Yacht, die Francis für sich selbst baute.

THE LARGEST SINGLE RIG IN THE WORLD AT THE TIME

When launched in 1981 the 28 metre yacht Diablesse with her sistership Must, the slightly smaller Chrismi II and La Concorde, Martin Francis had for a short time the worlds four largest sloops (a sailing yacht with one mast). The yachts where built in the south of France by Trehard and Chantier Navals de l'Esterel. Diablesse circumnavigated the world three times and won numerous regattas. Her lifting centerboard allows her access to shallow waters; up to eight guests can sleep in three cabins on board. The order for these yacht was a direct result of Prototype, a 14 metre long centerboard sloop that Francis built for himself.

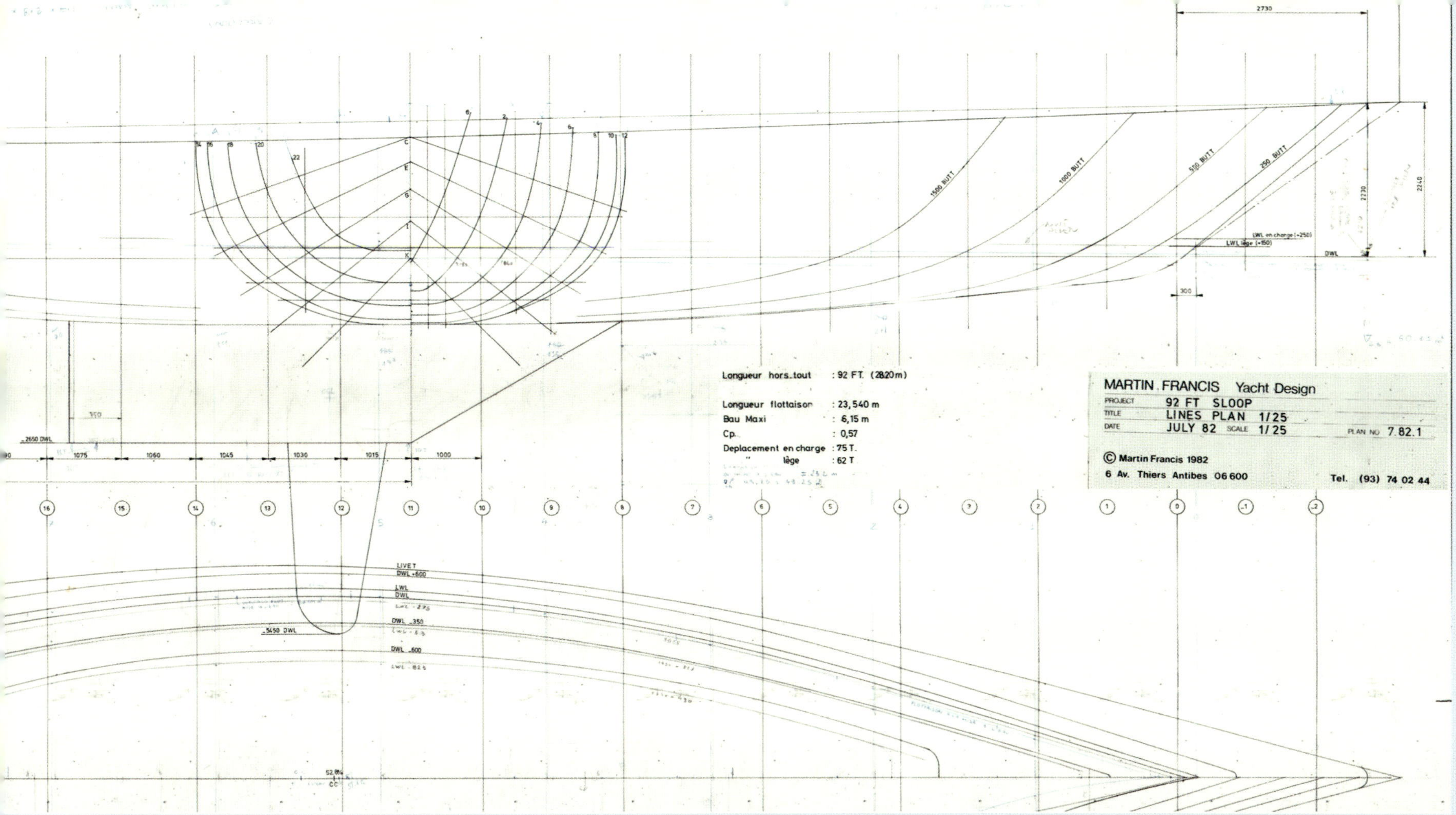

Longueur hors.tout : 92 FT. (28,20 m)
Longueur flottaison : 23,540 m
Bau Maxi : 6,15 m
Cp. : 0,57
Deplacement en charge : 75 T.
" lège : 62 T
MARTIN . FRANCIS Yacht Design
PROJECT 92 FT SLOOP
TITLE LINES PLAN 1/25
DATE JULY 82 SCALE 1/25 PLAN NO 7.82.1
© Martin Francis 1982
6 Av. Thiers Antibes 06 600 Tel. (93) 74 02 44

PROTOTYPE

VOM WELTUMSEGLER ZU EINEM VERKAUFSTOOL

Als die Yachtdesign-Geschäfte Ende der 1970er-Jahre etwas unrund liefen, schlug Francis' damalige Ehefrau Mimi vor, eine Pause einzulegen und mit den Kindern um die Welt zu segeln. Als sich für die Ansprüche der Familie kein geeignetes Boot fand, entschied Francis, selbst eines zu gestalten und im Anschluss auch zu bauen. Es entstand die 14 Meter lange PROTOTYPE aus Aluminium, die unglaublich schnell unterwegs war. Dass sie kein Interior besaß, wusste kaum jemand, ihre Performance beeindruckte indes mehrere Kunden, die daraufhin bei Francis ein Design in Auftrag gaben.

FROM A PROJECTED CIRCUMNAVIGATION TO A SALES TOOL

When architecture went through a recession at the end of the 1970s, Francis' then wife Mimi suggested they sailing around the world with their two young kids. When no suitable boat could be found for the family's needs, Francis decided to design one himself and build it as well. The 14 metres long PROTOTYPE was built in aluminium and on it's first trials was very fast. No one including Francis realised that this was because, with no interior, she was very light. However its performance impressed several customers, who then commissioned yachts from Francis so the round the world trip was abandoned.

MARTIN FRANCIS | DESIGN & INNOVATION

SHENANDOAH OF SARK

Yachten Yachts

EIN NEUES DECKSHAUS FÜR EINE EHRWÜRDIGE LADY

Die bereits im Jahr 1902 abgelieferte SHENANDOAH OF SARK gehört zu den ältesten und wohl auch schönsten Segelyachten der Welt. Der 54 Meter lange Dreimaster, der heute noch als Charteryacht unterwegs ist, hat eine extrem bewegte Geschichte und etliche Eignerwechsel hinter sich. Mitte der 1990er-Jahre stand für den Klassiker ein großes Refit bei McMullen & Wing in Neuseeland an, bei dem Martin Francis für SHENANDOAH OF SARK ein demontierbares Deckssalondach gestaltete – so konnte die Yacht komfortabel cruisen und auch Regatten segeln. Der Wiederaufbau der Yacht wurde preisgekrönt.

A NEW DECKHOUSE FOR A VENERABLE OLD LADY

SHENANDOAH OF SARK, delivered in 1902, is one of the oldest and most beautiful sailing yachts in the world. The 54 metres long three-masted schooner, which is still a successful charter yacht today, has an extremely eventful history and several changes of ownership behind her. She was once owned by Baron Bich (founder of BIC ball point pens). In the mid-1990s, the classic was restored at McMullen & Wing in New Zealand under the direction of Martin Francis who designed a removable deck saloon roof to enable SHENANDOAH OF SARK to both race and cruise in comfort, the restoration subsequently won an award.

*»Von den Woodstock-Machern
lernte ich eine wichtige Lektion:
Mit dem nötigen Willen und genügend Budget
kannst du eigentlich alles umsetzen.«*

“I learned an important lesson from the guys who organised Woodstock (the pop festival): With the necessary willpower and budget, you can actually do almost anything.”

Projekte
Projects

94 – 95

96 – 97

98 – 101

102 – 103

104 – 107

ULTIMX

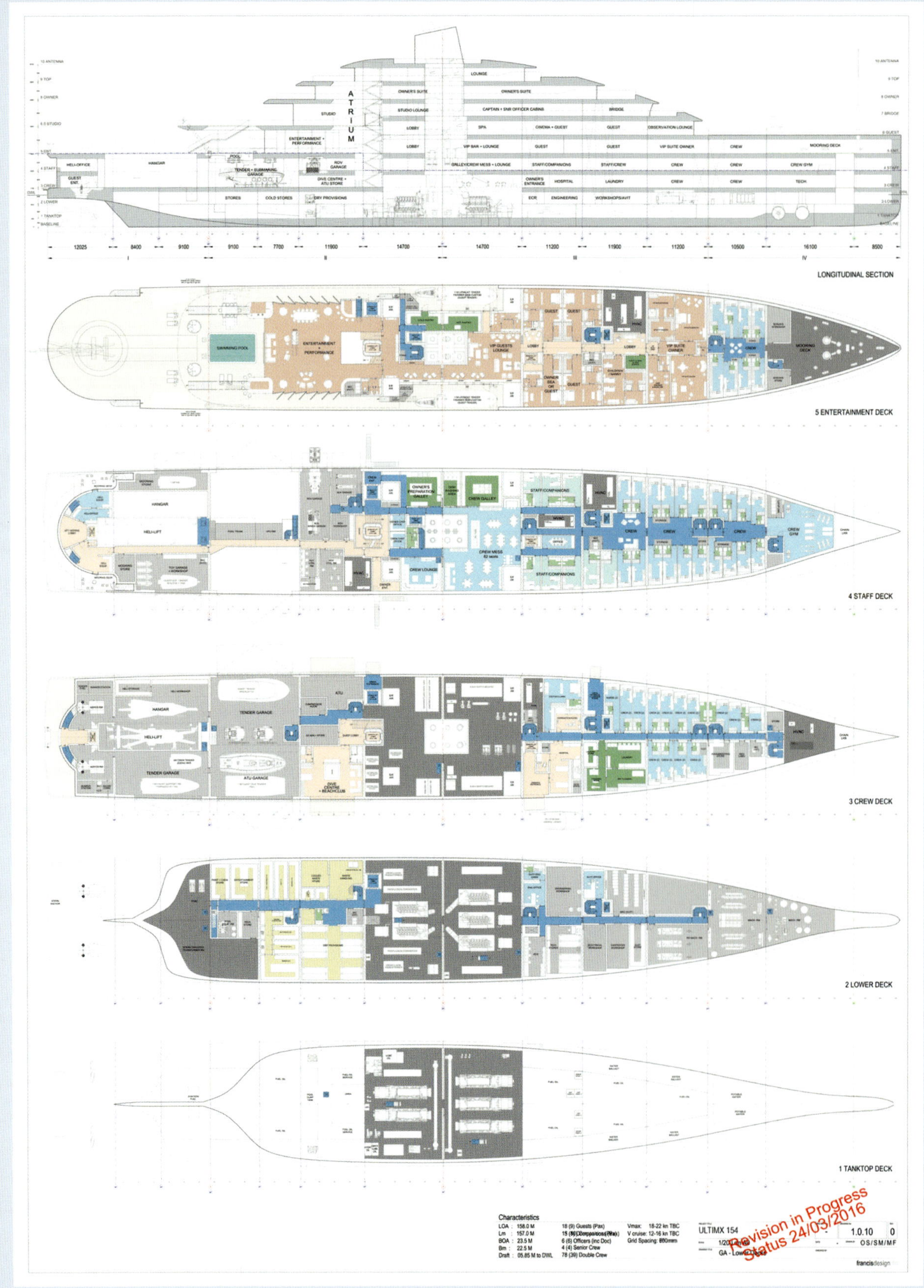
LONGITUDINAL SECTION
5 ENTERTAINMENT DECK
4 STAFF DECK
3 CREW DECK
2 LOWER DECK
1 TANKTOP DECK
Characteristics
LOA : 158.0 M
Lm : 157.0 M
BOA : 23.5 M
Bm : 22.5 M
Draft : 05.85 M to DWL
18 (9) Guests (Pax)
6 (6) Officers (inc Doc)
4 (4) Senior Crew
78 (39) Double Crew
Vmax: 18-22 kn TBC
V cruise: 12-16 kn TBC
ULTIMX 154
1.0.10
0
OS/SM/MF
Revision in Progress
Status 24/03/2016
francisdesign

UPGRADE FÜR EINEN ERFAHRENEN EIGNER

Dieses Projekt entwarf Francis für den erfahrenen Yachteigner und Philanthropen Paul Allen. Der Auftrag lautete, ein Schiff für weltweite Forschungs- und Entdeckungsfahrten zu entwerfen; an Bord sollte zudem ein Aufnahmestudio platziert werden – Rockmusik ist beziehungsweise war eine von Allens Leidenschaften. Die Yacht sollte weiterhin in der Lage sein, ein (natürlich gelbes) U-Boot für acht Personen aufzunehmen, ein ferngesteuertes Unterwasser-Fahrzeug und zwei Roboter. Außerdem forderte Allen zwei Hubschrauber, die geschützt in einem Hangar parken sollten.

AN UPGRADE FOR AN EXPERIENCED OWNER

This project was designed for the experienced yacht owner and philanthropist Paul Allen. The brief was to design a vessel for worldwide research and discovery operations, plus a recording studio and entertainment facilities for one of Allen's passions, rock music. To satisfy his love for underwater research and exploration the vessel should be capable of accommodating an eight-man (yellow of course) submarine called Pagoo, a remotely operated vehicle and two onboard robots. In addition he requested two helicopters protected in a hangar.

DAS ACHTERDECK ALS PARTY-PLATTFORM

Der Salon von ULTIMX sucht im Yachting seinesgleichen: Er ist 16 Meter breit, 21 Meter lang und 4,3 Meter hoch. Bodentiefe Flügeltüren zu den Seitendecks sorgen für eine natürliche Belüftung; sie sind nach achtern gerichtet, sehen wie Fischkiemen aus und lassen die Luft zirkulieren. In Kombination mit dem riesigen Achterdeck bietet der Salon ausreichend Volumen, um große Gesellschaften zu bewirten – unter anderem zu den Filmfestspielen von Cannes, während deren Paul Allen auf seine OCTOPUS immer zahlreiche Gäste einlud. Wie bei den Entwürfen von Martin Francis üblich befindet sich die Suite des Bauherrn über der Brücke.

AN AFT DECK FOR BIG PARTIES

On the main deck aft there is a vast entertainment space: 16 metres wide, 21 metres long and 4.3 metres high with full height hinged doors to the side decks facing aft like fish gills to allow natural ventilation and inside/outside living whenever possible. Combined with the enormous aft deck this provides space for entertaining large numbers of people for events such as the Cannes Film Festival where Paul Allen's party had become an annual must go to event. As is usual with Martin Francis' designs, the owner's suite is above the bridge to ensure maximum privacy and 180-degree views.

© francisdesign

DIE PERFEKTE LÄNGE? 154 METER

Das Eignerdeck ist so konzipiert, dass es immer privater wird, je mehr man sich nach vorne bewegt. Achtern hätte Allen seine privaten Gäste empfangen und unterhalten können, während es an Steuerbord eine Kunstgalerie gibt, der sich sein privates Arbeitszimmer anschließt, das zu seiner Suite samt Skylight führt. Sehr komfortabel ist der private Aufzug von der seitlichen Einstiegsklappe bis in die Eignersuite. Um diesem komplexen Auftrag gerecht zu werden, wurden verschiedene Yachtlängen zwischen 135 Meter und 157 Meter in Betracht gezogen. Letztlich wurden 154 Meter gewählt.

THE PERFECT LENGTH? 154 METRES

The owner's deck is designed to be progressively more private as one moves forward, aft is a reception area and outside deck where he can entertain guests, moving forward on the starboard side is an art gallery which leads through his private study to his forward-looking suite with an opening roof. Most importantly there is private lift from the side boarding platform directly to the owners suite. In order to meet this complex brief, several different lengths of vessel where considered ranging for 135 metres to 157 metres. In the final analysis the 154 design was selected as being best suited to the requirements.

MIT EINEM GLASAUFBAU DEM TREND VORAUS

Das 140 Meter lange Projekt Crystal Ball entwickelte Francis zusammen mit der Kieler Werft HDW. Die Yacht mit einem nahezu komplett verglasten Aufbau diente HDW als »Concept Yacht«, um mögliche Kunden für andere, etwas weniger radikale Aufträge anzusprechen. Bereits im Jahr 2007 erkannte Francis, was Jahre später zu immer größerer Nachfrage führte: Viele Eigner wünschen sich so viel Glas wie möglich auf einer Großyacht – zum einen um die Räume mit Licht zu fluten, zum anderen um eine stärkere Verbindung zwischen dem Interior und dem offenen Meer oder der Ankerbucht herzustellen.

A GLASS SUPERSTRUCTURE AHEAD OF THE TREND

Francis developed the 140 metre long Crystal Ball project together with the Kiel shipyard HDW. The yacht with an almost completely glazed superstructure served HDW as a "concept yacht" in order to address potential customers for other, somewhat less radical orders. As early as 2007, Francis recognised what led to increasing demand years later: Many owners wanted as much glass as possible on a large yacht – on the one hand to flood the rooms with light, and on the other to create a stronger connection between the interior and the open sea or anchorage.

CRYSTAL BALL

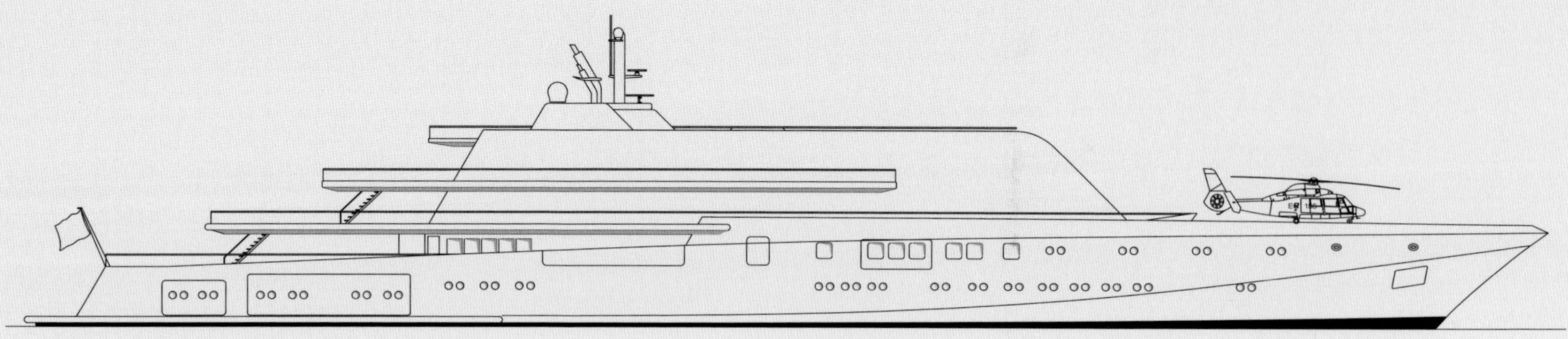

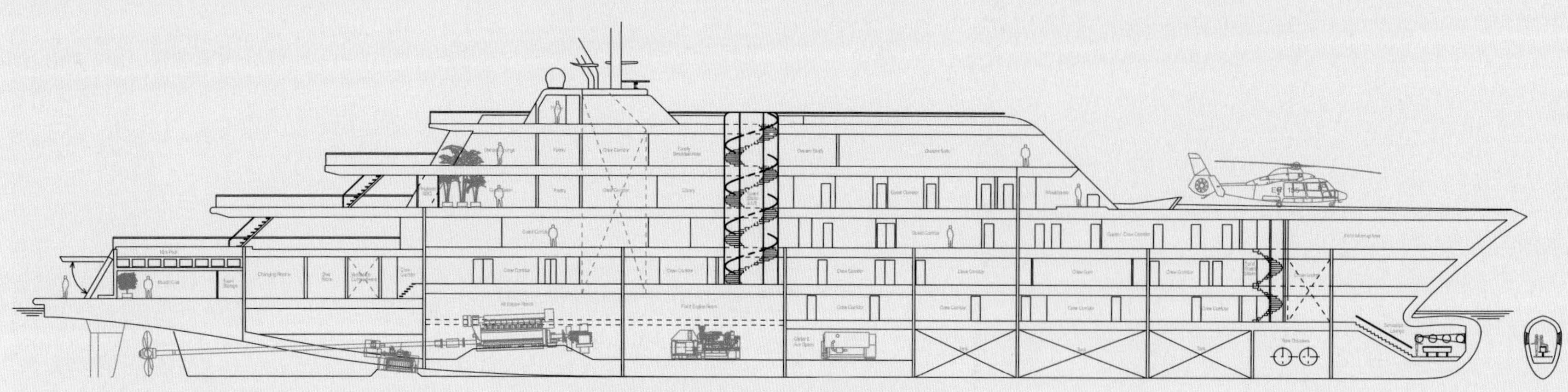

EIN ZEHN-METER-POOL UND EINE OBSERVATION LOUNGE

Im Heck des Hauptdecks, unterhalb der riesigen Terrassen (siehe vorige Doppelseite), besitzt die dieselelektrisch angetriebene CRYSTAL BALL einen zehn Meter langen Pool; ihn flankieren zahlreiche Außenmöbel. Wie schon bei ECO platzierte Francis die Eignersuite über der Brücke; dem Schlafzimmer schließen sich auf derselben Ebene ein Office, ein Frühstücksraum und ein Eignersalon an. Auf dem Tankdeck, in Höhe der Motoren und ganz vorn im Bugwulst, besitzt CRYSTAL BALL eine Observation Lounge. Durch Spezialscheiben können Gäste in Fahrt die Unterwasserwelt beobachten.

A TEN-METRE POOL AND AN OBSERVATION LOUNGE

In the stern of the main deck, below the huge terraces (see previous double page), the diesel-electric CRYSTAL BALL has a ten-metre-long pool, flanked by numerous outdoor furniture. This space is used to transport large yacht tenders during transatlantic voyages. As with ECO, Francis placed the owner's suite above the bridge; the bedroom is joined on the same level by an office, a breakfast room and an owner's salon. CRYSTAL BALL has an underwater observation lounge built into the bulbous bow to watch the dolphins and other marine life.

54 M SAILING KAT

RADIKALES STYLING AUF ZWEI RÜMPFEN

Katamarane, also Yachten auf zwei Rümpfen, stehen bei den UHNWIs dieser Welt eigentlich nicht so hoch im Kurs. Und doch beauftragte ein Interessent Martin Francis und German Frers mit der Gestaltung dieses 54 Meter langen Formats, das bei seiner Realisierung der längste Segelkatamaran gewesen wäre. Das Gewicht des Brandschutzes, das für eine Charteryacht dieser Größenordnung erforderlich gewesen wäre, hätten die Geschwindigkeitsvorteile des Kats gegenüber Einrümpfern zunichte gemacht. Das Projekt wurde deshalb nie realisiert.

RADICAL STYLING ON TWO HULLS

Catamarans, i. e. yachts on two hulls, are not so popular with the UHNWI's of this world. And yet a potential client commissioned Martin Francis to work with German Frers on the design of this 54-metre yacht, it would have been the longest sailing catamaran when it was built however owing to the weight of the fire protection required for charter vessels the potential speed advantages of a catamaran could not be achieved so this stealth machine was never built.

70/70

EIN AGGRESSIVES BEGLEITBOOT MIT RACING-RUMPF

Dieses Projekt heißt nicht ohne Grund »70/70«: Es ist 70 Fuß lang und erreicht als Höchstgeschwindigkeit 70 Knoten, umgerechnet auf Straßenverhältnisse also fast 130 km/h. Francis entwickelte diesen 24-Meter-Maxitender für einen erfahrenen Eigner, der sich jedoch nicht zum Bau durchringen konnte. Der Carbonrumpf der 70/70 wurde zusammen mit Fabio Buzzi entwickelt, einem der besten Konstrukteure für extrem schnelle Motorboote, für den Antrieb sollten zwei Gasturbinen sorgen. Übernachten können vier Gäste in zwei Kabinen, zwei Crewmitglieder finden in einer weiteren Kabine Platz.

AN AGGRESSIVE CHASEBOAT WITH A RACING HULL

This project is not called "70/70" without reason: It is 70 feet long and a projected maximum speed of 70 knots (almost 130 km/h in conventional terms). Francis developed this 24-metre maximum tender for an experienced owner, who in the end decided not to go so fast. The carbon fiber hull of the 70/70 was designed in conjunction with Fabio Buzzi, one of the best designers of extremely fast motorboats, with two gas turbines for the propulsion. Four guests can stay overnight in two cabins, two crew members can be accommodated in the forepeak.

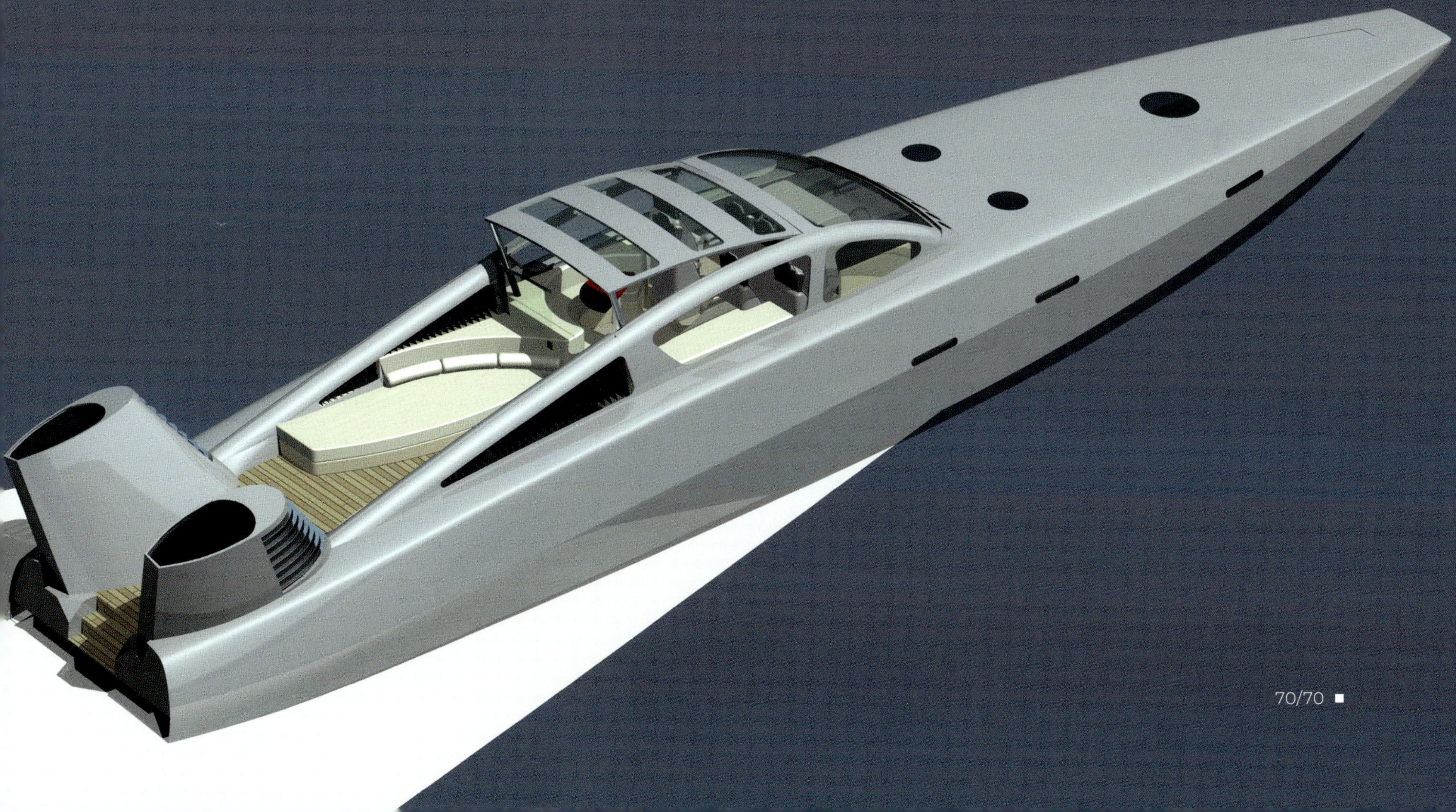

500 GT EXPLORER

DAS KONZEPT FÜR EIGNER AUS DER GENERATION Y

Werften, Designer und Broker müssen sich derzeit zwangsläufig mit einer neuen Generation von Eignern beschäftigen. Die 40-jährigen Erben, Silicon-Valley-HNWIs und Start-up-Millionäre entdecken das Yachting für sich, haben aber andere Vorstellungen als die Generation davor. Jüngere Eigner möchten die Welt entdecken, neue Reviere befahren und dabei möglichst viel erleben. Mit diesem 500 GT Explorer trägt Martin Francis diesem Trend Rechnung. So ist das Heck des 8,75 Meter schlanken Alubaus für zwei Tender reserviert, mit dem sich vom Mutterschiff aus Exkursionen unternehmen lassen.

A CONCEPT FOR OWNERS FROM GENERATION Y

Shipyards, designers and brokers are inevitably having to deal with a new generation of owners. The 40-year-old heirs, Silicon Valley HNWIs and start-up millionaires are discovering yachting for themselves, but have different ideas than the generation before them. Younger owners want to discover the world, sail new waters and experience as much as possible. Martin Francis is responding to this trend with this 500 GT Explorer. The stern of the 8.75 metre slim aluminium construction is reserved for two tenders, which can be used for excursions from the mother ship.

500 GT Explorer ►

VERBINDUNG ZWISCHEN INNEN UND AUSSEN

Der 50 Meter lange Explorer mit einer Verdrängung von 315 Tonnen und einem Tiefgang von 2,50 Metern ist momentan bei der französischen Werft JFA gelistet. Das hier gezeigte Interior ist ein moderner Vorschlag von Francis und Sandrine Melot für die angepeilte Zielgruppe. Große Türen und Fenster schaffen eine Verbindung zwischen den Räumen und der äußeren Umgebung. Auf dem schnittigen 500 GT Explorer mit eher sparsamen Aufbauten sollen sich Eigner und Gäste der Natur nahe fühlen. Ein spezielles Feature: Der 500 GT Explorer kann einen komplett geriggten Daysailer an Bord nehmen.

CONNECTION BETWEEN INSIDE AND OUTSIDE

The 50 metres long explorer with a displacement of 315 tons and a draught of 2.50 metres is currently listed at the French shipyard JFA, which builds its yachts on the French Atlantic coast. The interior shown here is a modern proposal by Sandrine Melot and Francis for the target group. Large doors and windows create a connection between the rooms and the outside environment. On the sleek 500 GT Explorer with rather economical superstructures, owners and guests should feel close to nature. One special feature of the 500 GT Explorer is it's ability to carry a large daysailer with the mast rigged.

SULTAN

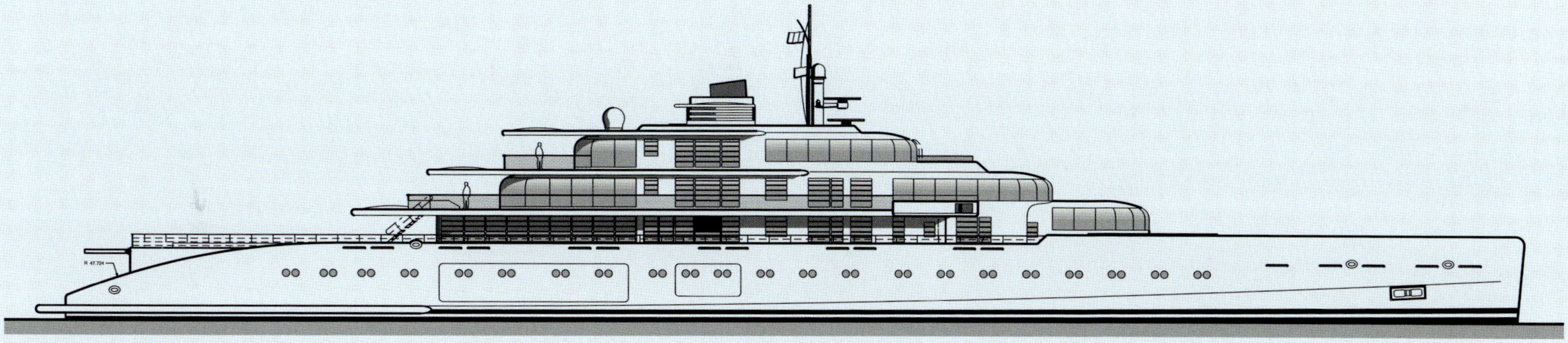

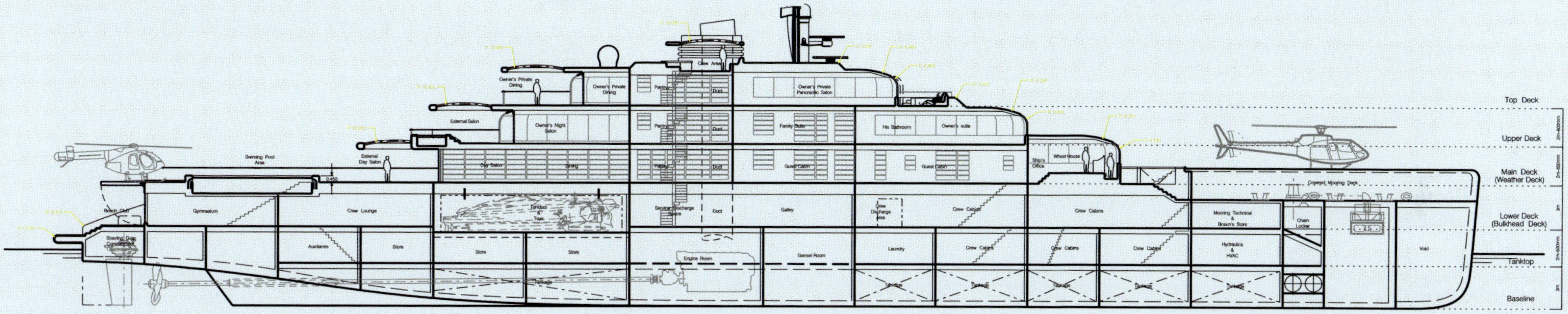

ÜBER DEN ATLANTIK MIT DURCHSCHNITTLICH 20 KNOTEN

Nachdem Francis für Larry Ellison KATANA (Ex-ECO) umgestaltet hatte, interessierte sich der Amerikaner für eine größere Yacht. Francis entwickelte daraufhin die 126 Meter lange SULTAN – ein mit 40 Knoten Höchstgeschwindigkeit sehr schnelles Format für seine Größe. Als Ellison seine neue Yacht woanders bestellte, interessierte sich ein türkischer Kunde für SULTAN. Ein Wirtschaftsskandal ließ den Auftrag dann leider platzen. Die 126-Meter-Yacht mit großflächiger Verglasung und einem sehr privaten Eignerbereich zählt zu Francis' Lieblingsprojekten.

ACROSS THE ATLANTIC WITH AN AVERAGE OF 20 KNOTS

After Francis had redesigned KATANA (ex-ECO) for Larry Ellison, the American became interested in a larger yacht. Francis then developed a 126-metre very fast concept vessel with a top speed of 40 knots. When Ellison decided to go elsewhere, a Turkish customer took over the project which he called SULTAN. Unfortunately, financial problems caused this second client to pull out. The 126-metre yacht with extensive glazing and a very private owner's area is still one of Francis' alltime favourite unbuilt projects.

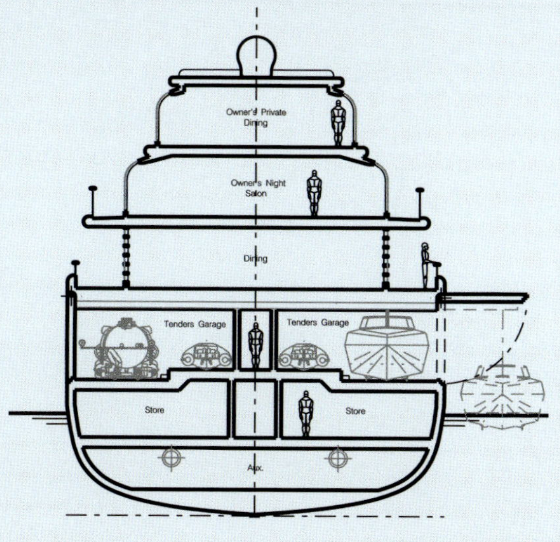

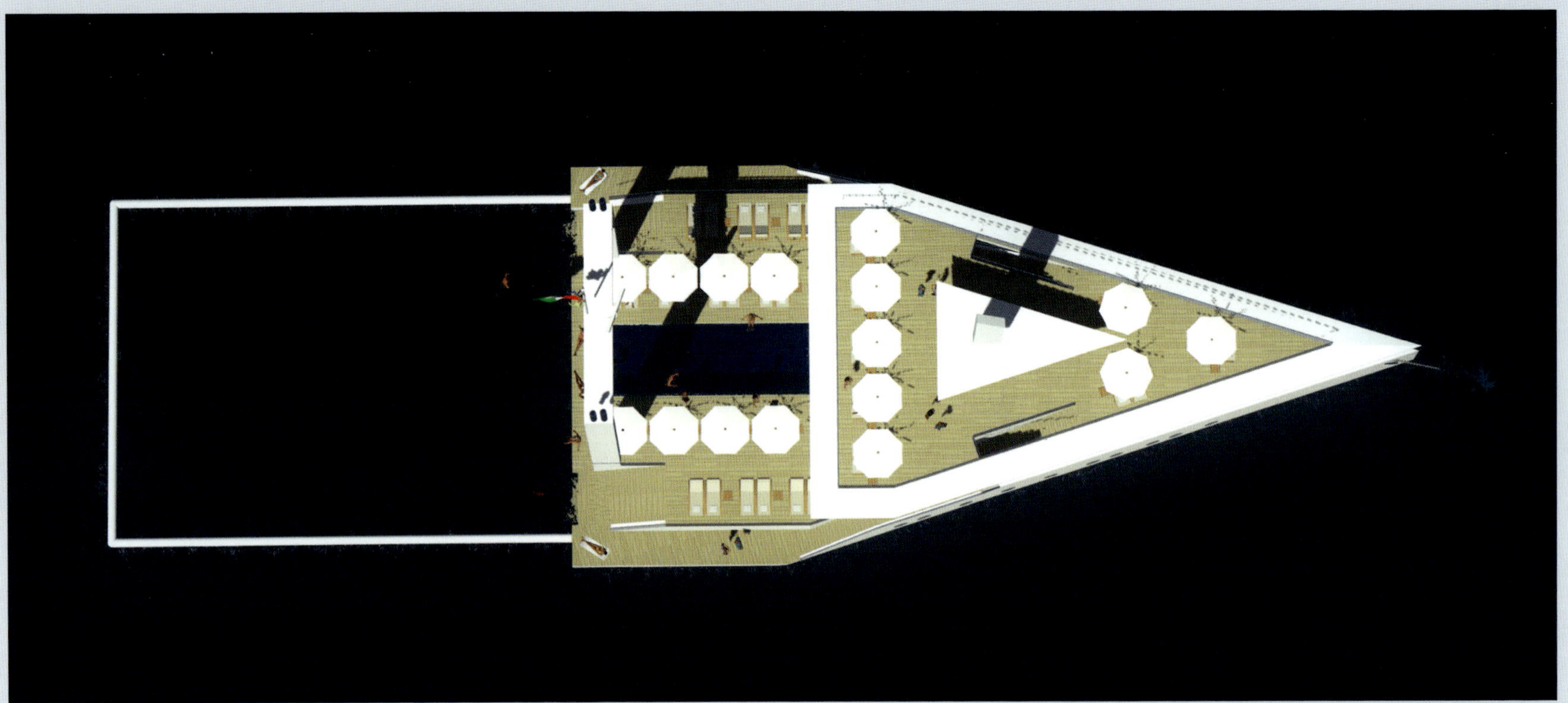

BEACH CLUB

WENN EIN HOTEL KEINEN STRAND BESITZT

Die Idee, einen schwimmenden Beachclub zu entwerfen, wurde von einem bekannten italienischen Hotel an Martin Francis herangetragen, das keinen Strandzugang besaß. Das Schiff ist rund 30 Meter lang und besitzt im Heck eine breite, flach abfallende Treppe. Gäste können so bequem schwimmen gehen und müssen keine Quallen oder Haie beim Baden fürchten – unter ihnen befindet sich ein großes Netz, das etwa 20 Meter lang und zehn Meter breit ist.

IF A HOTEL DOES NOT HAVE A BEACH

The idea of designing a floating beach club was presented by Martin Francis to a well-known Italian hotel who lacked a beach for its guests. The vessel is about 30 metres long and has a wide, flat staircase in the stern, which reaches into the water. Guests can swim comfortably in either the pool or the sea protected by a net from jelly fish or sharks. The mobile pool is approximately 20 metres long and ten metres wide.

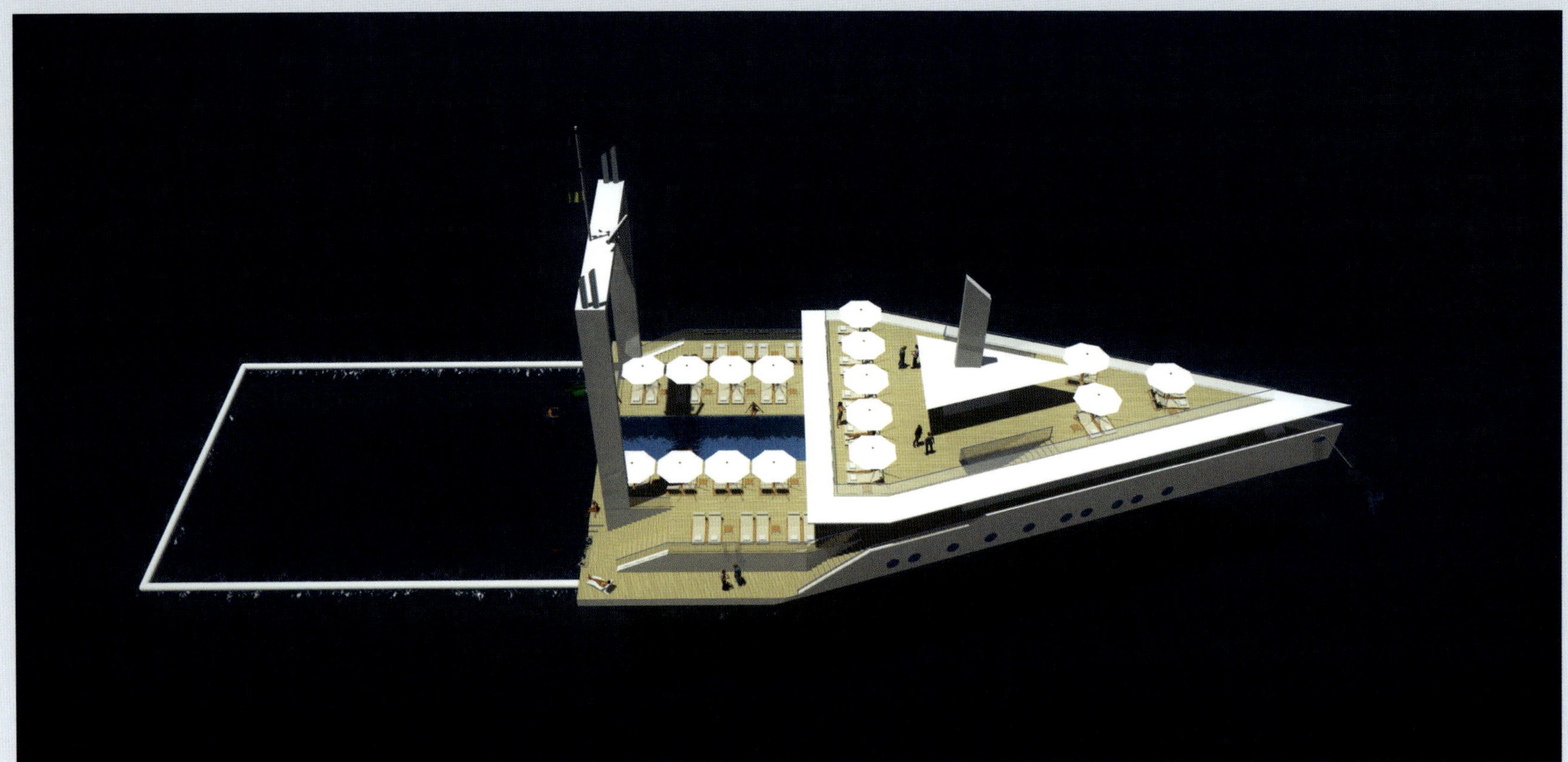

EIN RESTAURANT, EIN POOL UND EIN SPA

Die Baukosten für den Beachclub wären überschaubar: Martin Francis kalkulierte das Projekt mit einem Baubudget von knapp drei Millionen Euro. »Es ist ein simpler Stahlrumpf, und auch die zu installierende Technik ist nicht sonderlich komplex«, sagt er. Generatoren sorgen für die Stromversorgung, kleine Motoren für einen Topspeed von etwa fünf Knoten. So kann sich der Beachclub von Bucht zu Bucht bewegen und den Gästen eine wechselnde Szenerie bieten. Während auf dem Hauptdeck Restaurant, Pool und Bar angeordnet sind, kommen Wellness-Fans auf dem Unterdeck auf ihre Kosten. Dort hätte Martin Francis bei der Projektrealisierung ein Spa installiert.

A RESTAURANT, A POOL AND A SPA

In spite of the high standard of equipment for a floating beach club, the costs for such a project would be quite manageable. Martin Francis calculated the project with a construction budget of almost three million euros. "It is a simple steel hull and the technology to be installed is not particularly complex either," he says. Generators provide the power supply, small motors a top speed of about five knots. This allows the beach club to move from bay to bay and offer guests a changing scenery. While the restaurant, pool and bar are located on the main deck, wellness fans will get their money's worth on the lower deck.

»Wer wie ich oft von Selbstzweifeln geplagt wird, muss sich wohl in seiner Haut fühlen, um erfolgreich zu sein (was ich nicht wirklich bin).«

"As one who is often filled with self doubt, I think that you have to feel good in your skin to be really successful (which I am not)."

Architektur
Architecture

112 – 115

122

116 – 117

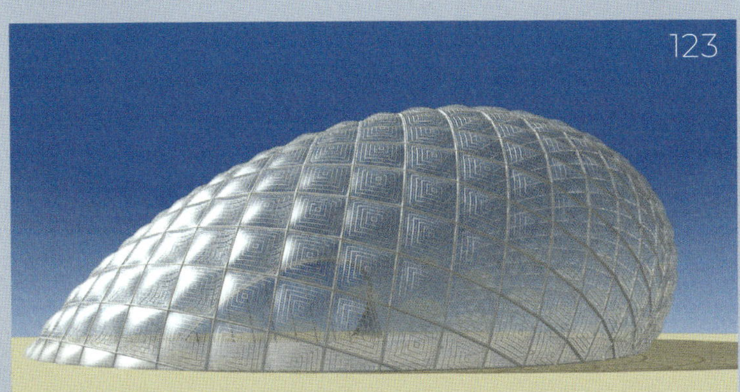
123

118 – 119

124 – 125

120 – 121

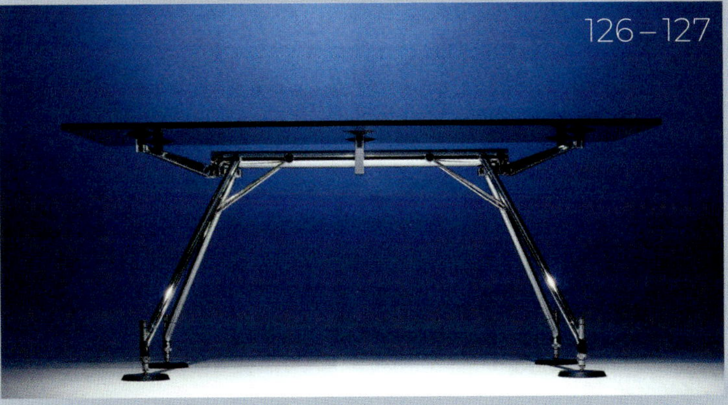
126 – 127

VITSŒ

DER CRYSTAL PALACE ALS REFERENZGEBÄUDE

Eines Tages, nachdem Martin Francis 30 Jahre lang nichts von einem alten Freund gehört hatte, bekam er von diesem einen Anruf. »Ich glaube, ich habe einen Job für dich.« Einige Tage später wurde er Mark Adams vorgestellt, CEO von Vitsœ, der mit seinem Unternehmen die Möbelsysteme von Dieter Rams herstellte. Adams suchte einen Designer, der ihm bei der Gestaltung seiner neuen Unternehmenszentrale half. Mit seinem Briefing »Zwölf-Meter-Yachten, Foster-Gebäude, Crystal Palace« hatte er bisher nicht den idealen Kandidaten gefunden. Francis hingegen wusste mit der Vorgabe sofort etwas anzufangen und begann bereits tags darauf mit der Umsetzung.

THE REFERENCE BUILDING WAS THE CRYSTAL PALACE

One day, having not heard from an old friend for 30 years, Martin Francis was surprised to have a phone call saying, "I think I have a job for you". A few days later he was introduced to Mark Adams the CEO of Vitsœ, the maker of Dieter Rams designed furniture who was looking for a designer to help him create his new headquarters. Adams had been struggling for many months to find a suitable architect, the brief mentioned 12-metre yachts, a building by Foster and the Crystal Palace. This ticked a lot of boxes for both Francis and Adams and they started to work the next day.

17

VIEL RAUM, LICHT UND EINE MUSEALE QUALITÄT

Das Vitsœ-Gebäude im englischen Leamington Spa setzt mit seiner Bauweise neue umweltfreundliche Maßstäbe. Es ist 135 Meter lang, 25 Meter breit und beherbergt Produktion, Lagerung, Verwaltung, Showroom und sogar Unterkünfte unter einem Dach. Die Konstruktion mit Pollmeier-Rahmen und Holzplatten kann als eine logische Erweiterung des von Dieter Rams entworfenen Vitsœ-Regalsystems bezeichnet werden. Flexibilität war ein Schlüsselkriterium, da Adams ein Gebäude wollte, das zukünftiges Wachstum unterstützt. Francis: »Mithilfe der inneren Säulen können Zwischengeschosse hinzugefügt werden, um zusätzliche Räume zu schaffen. Sein Licht verleiht dem Gebäude in gewisser Weise eine museale Qualität.«

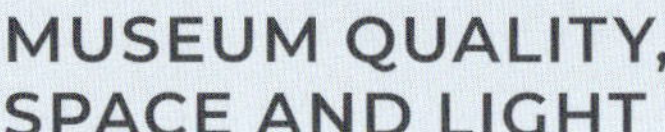

MUSEUM QUALITY, SPACE AND LIGHT

Natural light, timber frame and low carbon, the building sets new environmentally friendly standards. The Vitsœ building in Leamington Spa, England accommodates production, storage, administration, exhibition and even accommodation all under one roof 135 metres long and 25 metres wide. The Pollmeier laminated beach frame and cross laminated timber panels are an extension of the logic behind the Rams designed shelving system made by Vitsœ. Flexibility was a key criterion as Adams wanted a building that would support future growth. "Mezzanines can be added to create additional spaces, using the inner columns, whilst north lights give a museum quality to the building," says Francis.

16

LA VILLETTE

WEGWEISENDE GLASFASSADEN

Als Ingenieur des Pariser Centre Pompidou wurde Peter Rice eingeladen, den Architekten Adrien Fainsilber bei den besonderen Bauteilen des Nationalmuseums für Wissenschaft und Technik in Paris zu unterstützen. Fainsilber plante eine ähnliche Glasarchitektur, wie es Norman Foster bereits vorgemacht hatte, sodass Rice seinen Freund Martin Francis anrief. Die Herausforderung bestand darin, Gewächshausfassaden mit maximaler Transparenz zu schaffen. Als Francis darauf hinwies, dass man einfach verglaste Sicherheitsscheiben nehmen sollte, erfand Rice ein dazu passendes Kabelsystem. Dann entwickelten die beiden ihre patentierte, heute weltweit verwendete Bolzenbefestigung.

PIONEERING GLASS FACADES

Having been the engineer of the Centre Pompidou in Paris, Peter Rice was invited to assist architect Adrien Fainsilber with the special structures of the National Museum of Science and Technology, Fainsilber wanted to do glass like Foster so Rice called his friend Martin Francis and they founded the engineering firm RFR. The challenge was to create "serres" glasshouse type facades, with maximum transparency. When Francis pointed out that toughened glass was a flexible material, Rice had a eureka moment and invented the cable truss, they then developed the patented and articulated bolt fixing now used throughout the world.

LOUVRE

EIN YACHTRIGG FÜR DEN LOUVRE IN PARIS

Das Verglasungssystem des nationalen Wissenschaftsmuseums beeinflusste auch den weltberühmten Architekten I. M. Pei, der von François Mitterrand mit der Neugestaltung des Louvre-Museums beauftragt worden war. Peis Vorschlag einer Glaspyramide für den Haupteingang war äußerst umstritten, sodass er eine Simulation in voller Größe durchführen wollte. Francis hatte die Idee, mit einem Kran ein Kabelnetz in Form der Pyramide auf der Baustelle aufzuhängen. »Brillant«, sagte Pei, als das Projekt genehmigt wurde. Im Anschluss daran wurde RFR um Hilfe bei der Planung des komplexen Kabelsystems für die Glashalterung der Pyramide gebeten. Francis schlug vor, ein System analog dem auf Regattayachten zu nehmen, und rief seinen Freund und Experten Tim Eliassen bei Navtec an. Das System funktionierte, und Eliassen gründete daraufhin ein neues Unternehmen – TriPyramid Structures.

YACHT RIGGING FOR THE LOUVRE IN PARIS

The success of the Science Museum glazing system came to the notice of I. M. Pei who had been commissioned by President Mitterrand to revitalise the Louvre museum. Pei's proposal of a glass pyramid for the main entrance was extremely controversial so he wanted to undertake a full-size simulation. Francis came up with the idea of using a crane, out of view, to suspend a net of cables in the form of the pyramid on the site. "Brilliant," said Pei when the project was approved by Chirac, then the mayor of Paris. Following this RFR was asked to help with the engineering of the pyramid's complex cable support system. Francis suggested that the system be changed to that used on racing yachts and called his friend Tim Eliassen at Navtec. Tim got the job and formed a new company, TriPyramid Structures now world leaders in architectural hardware.

GLASPOOL

BITTE FALLEN SIE NICHT AUS DEM POOL!

Es sind nicht die großen Yachten auf Martin Francis' Website, die die meisten Klicks bekommen – es ist dieser Pool. In einem Garten gelegen, an dessen Gestaltung Francis in Südfrankreich mitgewirkt hat, war es sein Ziel, einen Pool zu schaffen, der bei Nichtnutzung gut aussah und von dem aus man die Aussicht genießen würde. Die 50 Zentimeter lange, freitragende Glasumrandung löst beide Aspekte, das einzige Problem ist laut Francis: »Man muss vorsichtig mit Kindern sein, damit sie nicht aus dem Pool herausfallen!«

BE CAREFUL NOT TO FALL OUT OF THE POOL!

It is not the yachts on Martin Francis' website that get the most hits but it is this pool.
Located in a garden Francis helped design in the South of France his aim was the create a pool that looked good when not in use and from which one could enjoy the view rather than the tiled edge. The 50-centimetre cantilevered glass edge solves both problems, the only problem says Francis, "You have to be careful with kids, that they don't fall out of the pool!"

HALLEY VI

FORSCHUNGSSTATION FÜR DIE ANTARKTIS

Im Jahr 2004 schrieben das Royal Institute of British Architects und der British Antarctic Survey einen Designwettbewerb für die Forschungsstation Halley VI in der Antarktis aus. Die Station sollte auf dem 150 Meter dicken und schwimmenden Brunt-Schelfeis stehen, wo extreme Wetterbedingungen herrschen und das nur wenige Monate im Jahr zugänglich ist. Da die Station aufgrund der Bewegung des Eises verlagerungsfähig sein musste, sahen Francis und sein Team Parallelen zum Yachting. Sie entwickelten selbsttragende, mobile Einheiten und wurden aus 86 Einsendungen unter die besten sechs gewählt. Zu den Teilnehmern gehörten einige der berühmtesten Namen der Architektur.

RESEARCH STATION FOR ANTARCTICA

In 2004, the Royal Institute of British Architects and the British Antarctic Survey announced a design competition for the Halley VI research station in Antarctica. To be situated on the 150 metres thick floating Brunt Ice Shelf, where extreme weather conditions prevail and which is only accessible for a few months of the year, the base has to be capable of relocation as the ice moves. Francis and his team saw parallels to yachting and developed self-sustaining, mobile units and out of 86 entries they were shortlisted among the best six submissions from some of the most illustrious names in architecture.

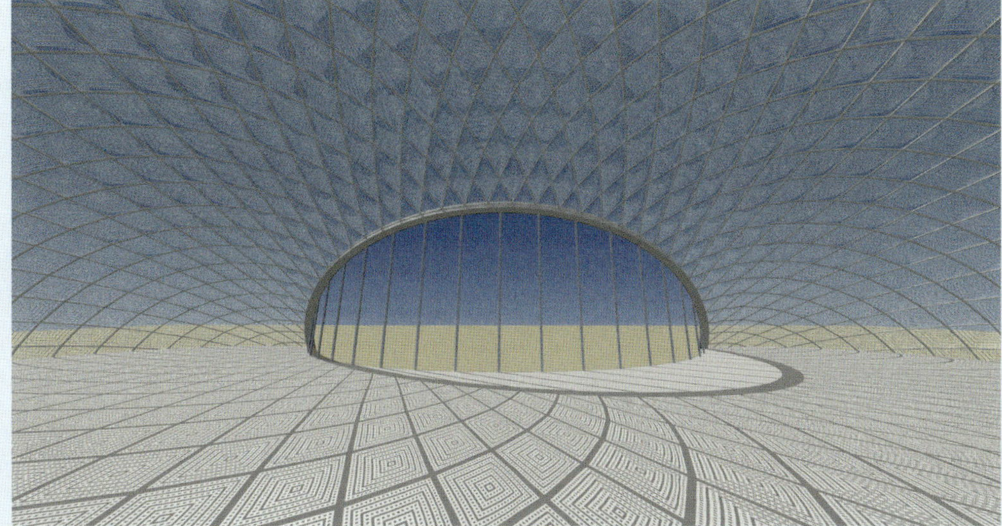

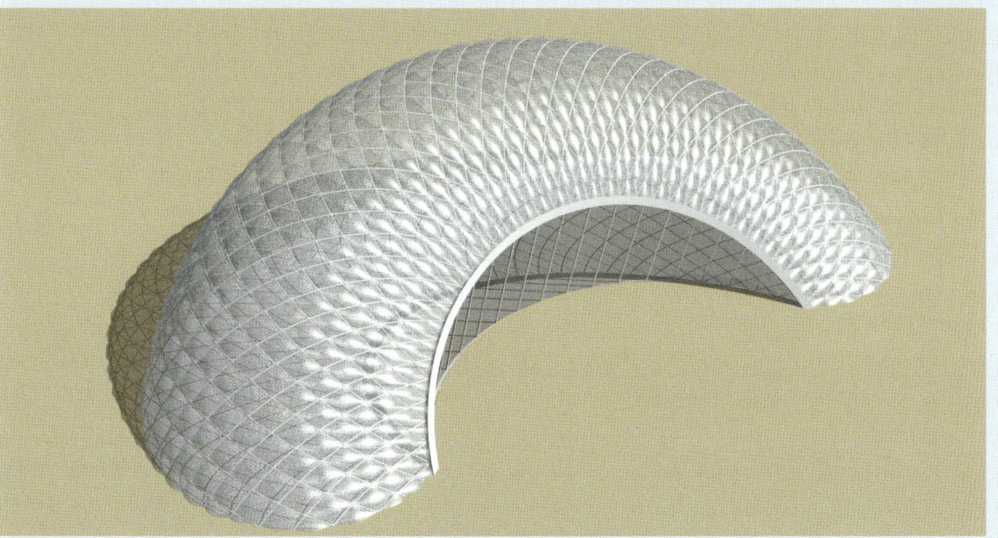

AIR PAVILLON

EINE TEMPORÄRE HALLE FÜR EINE GROSSE HOCHZEIT

Vor einigen Jahren erhielt Martin Francis einen äußerst interessanten Auftrag aus Dubai. Dort finden oft große Hochzeiten mit vielen Gästen statt, die einzigartige Locations erfordern. Francis schlug einen aufblasbaren Pavillon mit einem Durchmesser von 50 Metern vor. Obwohl die Idee nicht realisiert wurde, schuf Francis ähnliche luftunterstützte Strukturen, wenn auch in kleineren Abmessungen, für ein weltberühmtes Radrennen.

A TEMPORARY HALL FOR A BIG WEDDING

Martin Francis received an extremely interesting order from Dubai a few years ago. There large weddings with many guests often take place, they require unique venues. Francis proposed an air supported pavilion with a diameter of 50 metres. Although the idea was not realised, Francis also created similar inflatable structures, albeit in smaller dimensions, for a world-famous bicycle race.

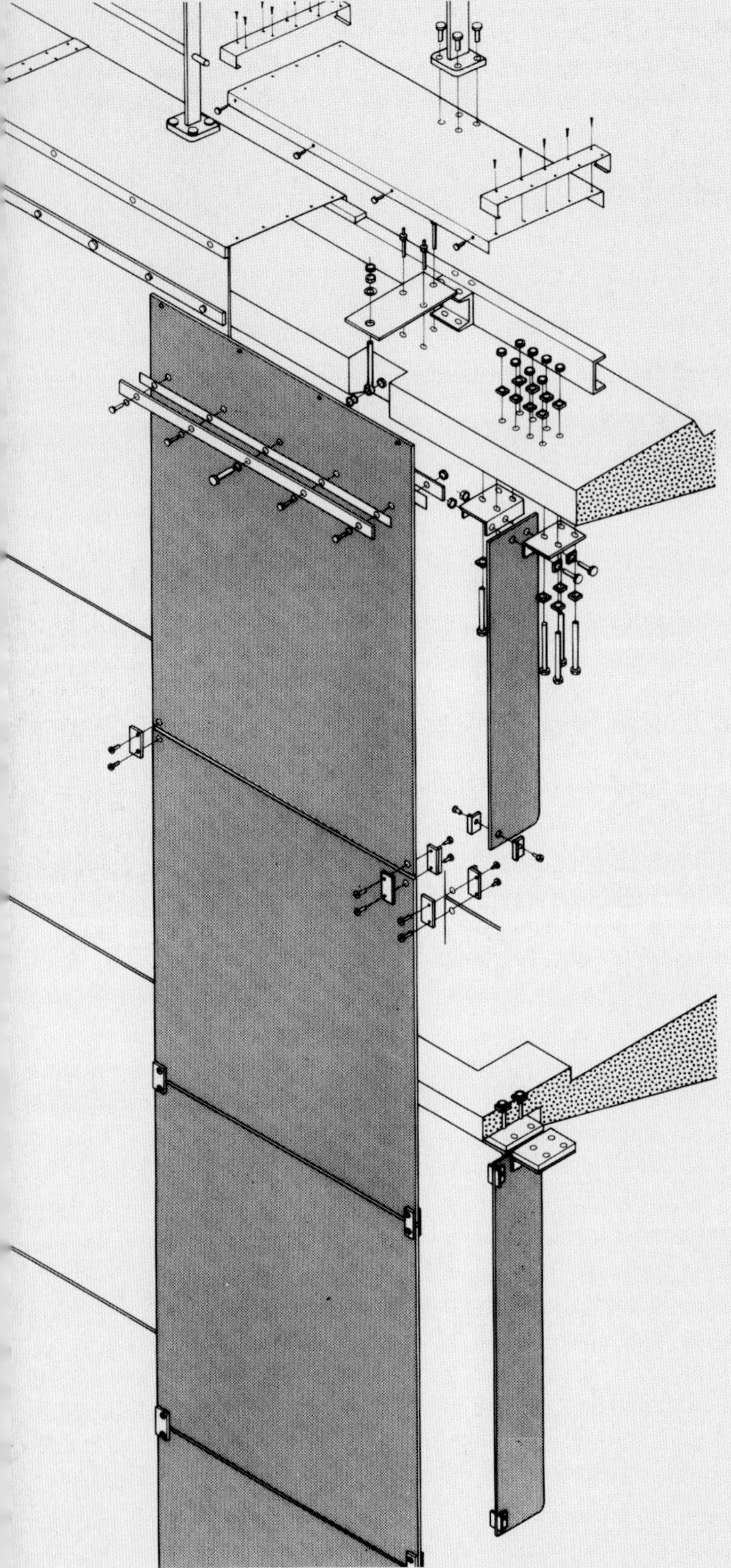

WILLIS FABER & DUMAS

DAS ERSTE GROSSE GLASPROJEKT

1970 erhielt Norman Foster seinen ersten Großauftrag für die Planung des Hauptsitzes von Willis Faber & Dumas in Ipswich; der Ingenieur dafür war Tony Hunt, Francis' alter Freund.

Das Projekt sollte mit einer durchgehenden Glaswand verkleidet werden. Foster hatte Probleme mit dem Design, also rief er Francis zur Hilfe. Das entwickelte Ergebnis war eine Glaswand, die quasi vom Dach hing. Das 1975 fertiggestellte Gebäude gilt als Pionierleistung auf dem Gebiet der Bürogebäude, wird jährlich immer noch von zahlreichen Architekturstudenten besucht und ist 1991 das jüngste Gebäude Großbritanniens, das den Status eines denkmalgeschützten Gebäudes der Klasse 1 erhielt.

THE FIRST MAJOR GLASS PROJECT

In 1970 Norman Foster was awarded his first major contract to design of the headquarters of Willis Faber & Dumas in Ipswich, the engineer was Tony Hunt, Francis' old friend.

The project was to be wrapped with a continuous glass wall and Foster was having problems with the design, so he called on Francis to help. The result was a glass wall suspended from the roof as shown on the illustration, drawn by hand, by Francis at the time. The building completed in 1975, is regarded as a pioneering achievement in the field of office buildings, is visited by numerous architecture students and in 1991 it became the youngest building in Britain to be given Grade 1 listed building status.

TECNO NOMOS

ERSTE MÖBEL UND EIN WELTBERÜHMTER TISCH

Erst im Internat und später im Central Saint Martins entwarf und fertigte Francis zahlreiche Möbelstücke, die sich noch heute in seinem Besitz befinden. Das berühmteste Stück, mit dem er in Verbindung gebracht wird, ist jedoch Norman Fosters »Tecno Nomos«-Tisch, für den Francis in seiner Zeit als Direktor von Foster Associates verantwortlich war. Der Nomos-Tisch war eine Entwicklung eines Modultisches, der für das Renault-Gebäude von Foster in Swindon entwickelt wurde. Er gilt als Ikone, wird immer noch verkauft und findet sich in unzähligen Büros, Wohnungen und Villen.

EARLY FURNITURE AND A WORLD-FAMOUS TABLE

When he was at boarding school and later at the Central Saint Martins, Francis designed and made numerous pieces of furniture which are still in his possession today. However, the most famous piece with which he is associated, is Norman Foster's "Tecno Nomos" table, for which Francis was responsible when he was a Director of Foster Associates. The Nomos table was a development of a 'lunar module' table designed for Foster's Renault building in Swindon. Considered an icon, it is still being sold and is found in countless offices, apartments and villas.

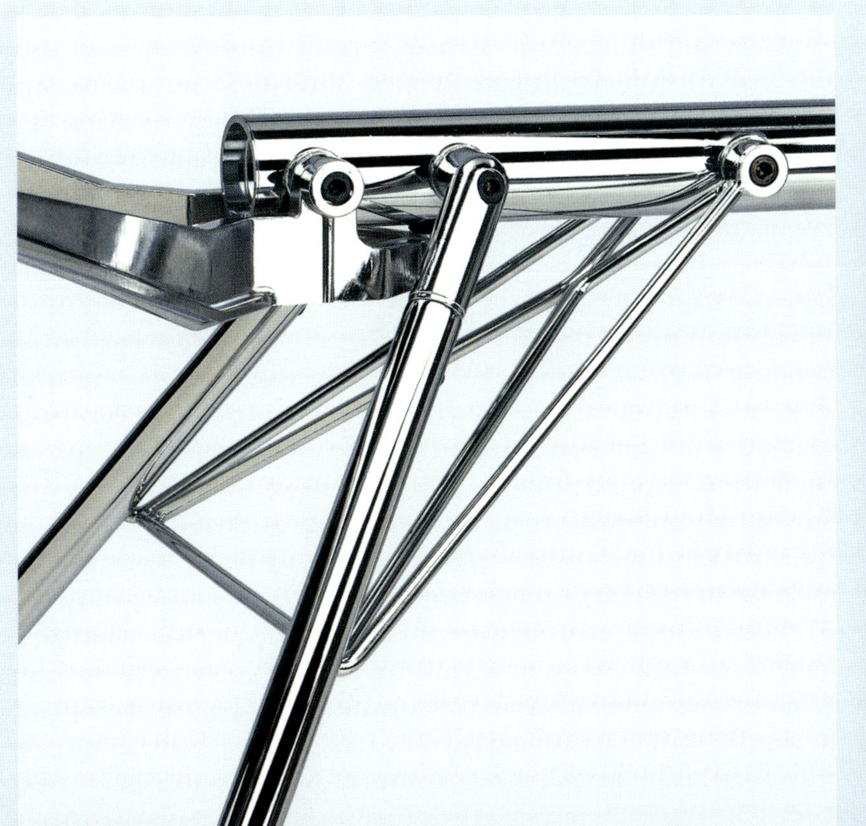

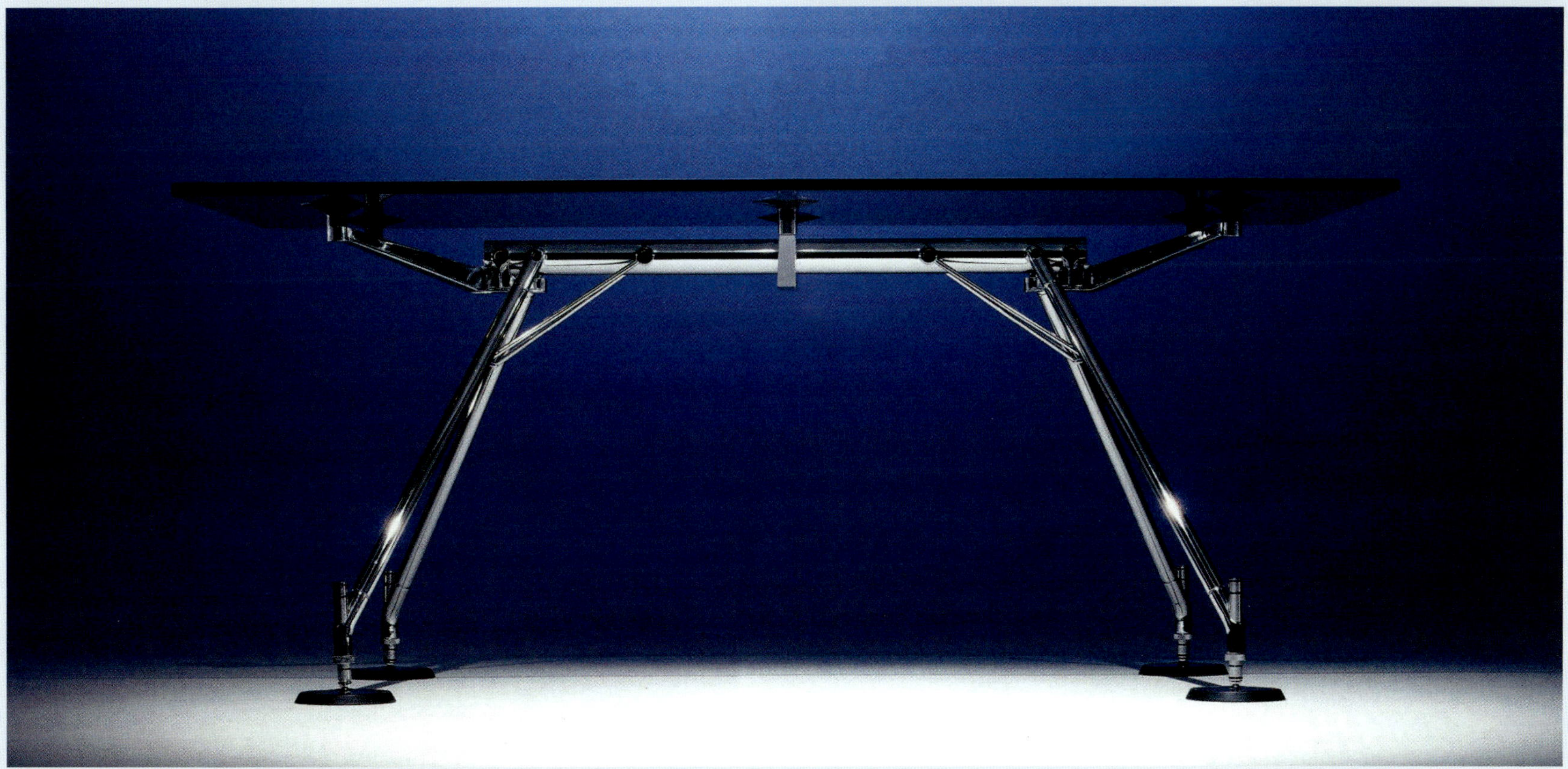

Kunst
Art

100 FAHNEN ZUM GEBURTSTAG

Wahrscheinlich durch seine Kindheit oder gar durch seine Gene – seine Mutter studierte schließlich bei Henry Moore – forcierte Martin Francis auch seine künstlerischen Ambitionen. Ein bemerkenswertes Projekt war etwa die Installation zum 100. Geburtstag des Vaters eines Freundes vor dessen englischem Landhaus. Francis positionierte dafür auf einem Rasenstück genau 100 Flaggen – als Symbol für das ehrwürdige Alter des »Geburtstagskindes«. Die Flaggenstöcke bestehen aus Angelruten, die Flaggen aus dünnem, rotem Spinnakertuch, die selbst bei kleinen Windstößen flattern und dann ein herrliches Bild abgeben.

100 FLAGS FOR A BIRTHDAY PARTY

Perhaps because of his upbringing, his mother studied with Henry Moore, Francis has always experimented in art. A remarkable project derived from his use of fishing rods made out of carbon fibre mobiles. This installation for the 100th birthday of the father of a friend at an English country home. Francis positioned a grid of exactly 100 flags on the surrounding lawn – as a symbol for the venerable age of the great man. The flagpoles consist of fishing rods, the flags, thin strips of red spinnaker cloth, flutter in the slightest breeze creating an undulating virtual surface.

▸

HOMMAGE AN FRANK STELLA

Als Folge seiner Zusammenarbeit mit Frank Stella (Seite 134/135) entwarf Francis diese Skulptur, die in einem weitläufigen Garten in Südfrankreich steht, nur wenige Kilometer von der Villa entfernt, in der Francis lebt. Im Zuge einer Umstrukturierung des Gartens installierte Francis dort für den Kunden das Kunstwerk, das er eine »Hommage an Frank« nennt. Sie besteht aus zwei stählernen, je 60 Millimeter starken Wänden mit 2,50 Meter Durchmesser, die durch das Aufschneiden in konzentrische Quadrate und Kreise beim Betrachten einen Unendlichkeitseffekt auslösen.

TRIBUTE TO FRANK STELLA

Following his collaboration with Frank Stella (see page 134/135), this sculpture is set in a large garden, a few kilometres from the villa where Martin Francis lives in southern France. As part of a remodeling of the garden, Francis installed the artwork, which he calls "Hommage to Frank", for the client. It consists of two 2.5 metre square 60 mm thick steel sheets, which, when cut into concentric squares and circles, create an infinity effect when viewed. The fact that the mild steel will eventually rust away is a deliberate reflection on ones mortality.

EINER DER GRÖSSTEN KÜNSTLER UNSERER ZEIT

Mit Frank Stella, einem der größten Künstler der Gegenwart, verbindet Martin Francis seit 1992 eine kollegiale Freundschaft. Nachdem sein Geschäftspartner Peter Rice gestorben war, führte Francis dessen Zusammenarbeit mit Stella fort, um ihn in puncto Materialien, Verarbeitung, 3-D-Druck und Visualisierung zu unterstützen. Zusammen entwickelten sie etwa »The Broken Jug«, eine 45-Tonnen-Skulptur, die auf der CMN-Werft in Cherbourg produziert wurde, sowie die Skulptur »Prinz Friedrich von Homburg, Ein Schauspiel, 3X« für die National Gallery in Washington. Francis assistierte Stella außerdem beim Design seiner Retrospektive »Frank Stella: Painting into Architecture« im Metropolitan Museum von New York im Jahr 2007.

ONE OF THE GREATEST ARTISTS OF OUR TIME

In his own words Martin Francis considers he has had the privilage of working with Frank Stella, one of today's greatest artists, since 1992. After his business partner Peter Rice died, Francis continued his collaboration with Stella assisiting him in the fabrication of monumental sculpture he also introduced him to parametric computer modelling and 3D printing. Together they developed "The Broken Jug", a 45-ton sculpture produced at the CMN shipyard in Cherbourg and built the sculpture "Prinz Friedrich von Homburg, Ein Schauspiel, 3X" at the National Gallery in Washington. Francis also assisted Stella with the design of "Frank Stella: Painting into Architecture" a major retrospective in the Metropolitan Muesuem in New York in 2007.

(*) actually that's not quite true. When Severinda is out a wall piece La Columba Ladra is in, as is clearly indicated in the drawings. cheers! F.

Dear Martin,

I hope this material is not as confusing as it first appears. I have supplied plans marking the location of objects and I have provided photo material of the objects. My hope is that you can produce 3D renderings of the objects in their intended home, i.e. the Met's roof garden + its Kimmelman gallery.

My roof plan has 5 white spots marking the placement of the sculptures which I have sent in various photo forms. You will have to put in the Brazilian Chinese Leaves structure to complete the arrangement.

I have sent 6 pieces of paper to represent my intentions for the Kimmelman gallery installation. They are exceptionally repetitious. The floor plan + the east + west wall elevations are repeated once, although I'm sure it will seem to be much more than that. The so-called square to curve partition called Severinda is the only difference(*). It is included in one version + excluded in the other.

So I'm hoping you can make 2 renditions of the Kimmelman gallery + 1 rendition of the roof garden. Thanks

F.

MARTIN FRANCIS | DESIGN & INNOVATION

Kreuzfahrtschiffe
Cruise liners

Kreuzfahrtschiffe Cruise liners

CELEBRITY SOLSTICE

EIN SCHIFF SETZT NEUE STANDARDS

Von Yachten zu Kreuzfahrtschiffen ist es nur ein kleiner Schritt, und so wurde Martin Francis von Royal Carribeans Harri Kulovaara mit der Gestaltung der CELEBRITY SOLSTICE beauftragt: 317 Meter lang, 36,80 Meter breit, ausgestattet mit 1.426 Gästekabinen und bewirtschaftet von einer 1.270-köpfigen Crew. Die CELEBRITY SOLSTICE wurde auf der Papenburger Meyer-Werft gebaut und war das erste von insgesamt fünf Schiffen, das neue Industriestandards setzte.

A SHIP SETS NEW STANDARDS

It is only a small step from yachts to cruise ships; and so Martin Francis was commissioned by Royal Caribbean's Executive VP of Maritime, Harri Kulovaara, to design the hull and exterior of the CELEBRITY SOLSTICE: 317 metres long, 36.8 metres wide (more than the panama canal) equipped with 1,426 guest cabins and managed by a crew of 1,270. CELEBRITY SOLSTICE built at the Meyer shipyard in Papenburg was the first of five ships which set new industry standards.

CELEBRITY EQUINOX

MIT ÜBER 67.000 KILOWATT AUF 24 KNOTEN TOPSPEED

Die Celebrity Equinox ist ein Schwesterschiff der Celebrity Solstice und wurde ebenfalls von der amerikanischen Reederei Celebrity Cruises bei der Papenburger Meyer-Werft in Auftrag gegeben. Ihr Bau begann im September 2007, von Celebrity Cruises übernommen wurde das 317-Meter-Schiff bereits im Juli 2009. Für den Antrieb der Celebrity Equinox sorgen Azipods von ABB. Jeder besitzt eine Leistung von 20,5 Megawatt, die das knapp 37 Meter breite Schiff auf einen Topspeed von 24 Knoten bringen. Für die Energie sorgen vier Wärtsilä-Motoren mit insgesamt 67.200 Kilowatt.

WITH OVER 67,000 KILOWATTS AT 24 KNOTS TOP SPEED

The Celebrity Equinox is a sister ship of the Celebrity Solstice and was also commissioned by Royal Caribbean's celebrity brand at the Meyer shipyard in Papenburg. Its construction began in September 2007, the 317-metre ship was deliverd to Celebrity Cruises in July 2009. Azipods from ABB are responsible for the propulsion of the Celebrity Equinox. Each has an output of 20.5 megawatts, bringing the ship, which is almost 37 metres wide, to a top speed of 24 knots. The energy is provided by four Wärtsilä motors with a total of 67,200 kilowatts.

Celebrity EQUINOX

CELEBRITY ECLIPSE

DER EFFIZIENTESTE KREUZFAHRER DER WELT

Die Nummer drei in der Celebrity-Baureihe bei der Papenburger Meyer-Werft war die CELEBRITY ECLIPSE. Sie wurde im April 2010 in Dienst gestellt und legt mit – im Vergleich zu heutigen Zahlen – noch bescheidenen 2.850 Passagieren ab. Die CELEBRITY ECLIPSE wurde im Jahr 2010 als das energieeffizienteste Kreuzfahrtschiff der Welt ausgezeichnet. Durch den Einsatz von Photovoltaik-Anlagen, eine optimierte Hydrodynamik und einen speziellen Unterwasseranstrich verbraucht sie 30 Prozent weniger Energie als andere Schiffe gleicher Art.

THE MOST EFFICIENT CRUISESHIP IN THE WORLD

The number three in the Celebrity series at the Meyer shipyard in Papenburg was the CELEBRITY ECLIPSE. It was put into service in April 2010 and has a modest 2,850 passengers compared to today's figures. In 2010, the CELEBRITY ECLIPSE was named the most energy-efficient cruise ship in the world. Through the use of photovoltaic systems, optimised hydrodynamics or a special underwater coating, it consumes 30 percent less energy than other ships of the same type.

Celebrity ECLIPSE

*»Es gibt zwei Arten von Menschen:
die, die Geld machen, und die,
die Dinge produzieren.
Die Produzenten der Dinge
machen sehr selten Geld.
Ich gehöre leider dazu.«*

“There are two sorts of people, those who make money and those who make things. People who make things rarely make money, I make things.”

Wegbegleiter
People

In den vergangenen 50 Jahren durfte ich mit außergewöhnlichen Persönlichkeiten zusammenarbeiten. Abgesehen von meiner Familie haben die folgenden Menschen mein Leben und meine Karriere nachhaltig beeinflusst.

Over the last 50 years, I have been privileged to know and work with some exceptional people, apart from my extended family, the following are some of the most important ones who have influenced my life and career.

Tony Hunt

Tony, mittlerweile ein sehr bekannter Ingenieur, ist ein Freund von mir, seitdem ich 15 Jahre alt war. Wir wuchsen zusammen auf, er war der Erste, der in meine Fähigkeiten vertraute, mir einen Auftrag gab, und er stellte mich Norman Foster vor, wofür ich ihm ewig dankbar bin. Während ich bei Tony arbeitete, kam Chip Monck in das Office und fragte, ob wir für die Rolling Stones einen demontierbaren Proszeniumbogen berechnen könnten. Als ich ihm bereits am nächsten Tag vor dem Büro eine Konstruktion aus Aluminium präsentierte, wurde ich auf einmal stellvertretender Produktionsleiter der 1970er-Stones-Tour.

Tony, a renowned engineer, has been a friend of mine since I was 15 years old, we where neighbours and when I was a cabinet maker, he gave me my first commission to build a plan chest. He was the first to trust my abilities, subsequently he offered me a job and introduced me to Norman Foster, for which I am eternally grateful. It was while I was working with him that Mr Chip Monck came into the office wanting somebody to calculate a demountable proscenium arch for the Rolling Stones, I came up with an idea using zip up aluminium scaffolding and demonstrated it the following day in front of the office. As a result of which, I became the assistant production manager of their 1970 European tour.

Lord Norman Foster

Heute ist er weltbekannt, viel bekannter als ich, und gehört sicher zu den Top-5-Architekten weltweit und war die wichtigste Person zu Beginn meiner Karriere. Als Foster allerdings seine Karriere startete, arbeiteten wir 1967 zu dritt aus einem Apartment in Hampstead heraus. Das hatte noch nichts mit dem gemein, wie das Büro Fosters heute aufgestellt ist.

In den folgenden 25 Jahren arbeitete ich mit Foster an verschiedenen Projekten, unter anderem am Carre d'Art in Nîmes, am Tecno-Möbelsystem und am Willis-Gebäude in Ipswich, das meinen Ruf als »Glass Man« formte.

Wie viele Kollegen wurde ich von Fosters zahlreichen Talenten beeinflusst. Er ist einfach ein guter Pitcher, ein Verkaufs- und Marketingtalent, ein Mann mit Sportsgeist. Es war ein Privileg, zu Beginn seiner Karriere mit ihm zu arbeiten, und ich freue mich einfach, dass wir immer noch befreundet sind.

The celebrated architect, undoubtedly one of the top five in the world, was the most important person at the start of my career in design. I started working with him in 1967 as one of three people working in a small Hampstead apartment, a far cry from the vast Foster + Partners organization today. On an off over the following 25 years I worked with Norman on several projects notably the Willis Faber glass wall, which established my reputation as the 'glass man', the Carre d'Art in Nîmes and the Tecno furniture system.

Like many of my generation, I was greatly influenced by Norman's many talents, draftsman, salesman, sportsman and aviator and consider it was a privilege to have been part of his team during the early years. I am happy to still count him among my friends 50 years later.

▸

Jean Prouvé

Le Corbusier sagte einmal: »In Jean Prouvé vereinigen sich Architekt und Ingenieur, richtiger noch, Architekt und Baumeister, denn alles, was er anfasst und gestaltet, bekommt sofort eine elegante und plastische Form, mit glänzend verwirklichten Lösungen in Bezug auf Haltbarkeit und industrielle Fertigung.« Unter anderem saß Prouvé der Jury des Centre Pompidou vor und war ein Vorbild für Renzo Piano, für Norman Foster, für Richard Rogers und auch für mich. Ich hatte das Vergnügen oder besser die Ehre, mit ihm Projekte zu diskutieren, die ich für Foster und Rogers umsetzte. Er war zudem ein wunderbar bescheidener Mann. Ich erinnere mich noch sehr gut an die Situation, als ich ihn in seinem Atelier in Paris besuchte, wo er mit einer Kehrschaufel auf Händen und Knien den Dreck auffegte – und das im Alter von 80 Jahren!

Le Corbusier once said: "Jean Prouvé unites architect and engineer, more correctly architect and builder, because everything he touches and designs immediately takes on an elegant and plastic form, with brilliantly realised solutions in terms of durability and industrial production." Among other things he was the president of the jury for the Centre Pompidou and a great inspiration for Richard Rogers, Renzo Piano and Norman Foster. However, a marvellously modest man, I remember visiting him in his studio in Paris to find him sweeping up on his hands and knees with a dustpan and brush aged 80! It was an honour, or rather a privilege, to have worked with him on projects I was doing at the time for both Richard and Norman.

Dr. Michael Papo

Eines Tages klingelte mein Telefon, und eine Stimme fragte: »Sind Sie Martin Thomas?« Ich verneinte, doch das schien den Mann nicht zu verunsichern. »Ok«, sagte er, »aber Sie sind trotzdem der, der in Südfrankreich schnelle Segelyachten entwirft?« Jetzt stimmte ich ihm zu und hörte noch: »Ich steige morgen in New York in die Concorde, dann sehen wir uns.« Dr. Papo, Jude mit jugoslawischen Wurzeln, war ein sehr erfolgreicher Geschäftsmann, der zu meinen ersten Yachtkunden überhaupt gehörte. Für ihn entwarf ich 1982 La Concorde (heute Dark Star), damals eine der größten Slups weltweit.

One day my phone rang, and a voice asked, "Are you Martin Thomas?" I said no, but that didn't seem to unsettle him. "Ok," he said, "but are you the man who designs large sailing yachts in the South of France?" Now I agreed with him, so he said: "I'll get on the Concorde tomorrow in New York and I'll come and see you." Dr Papo was a larger than life businessman with Yugoslavian Jewish roots, he was one of my very first yacht clients for whom I designed the 26-metre sloop La Concorde in 1982 (now called Dark Star) then one of the largest sloops in the world.

Peter Rice

Peter war ein Genie, ein Mentor für etliche Architekten und der vielleicht beste Ingenieur des 20. Jahrhunderts – zumindest empfinden ich und viele meiner Kollegen es so. Er war beispielsweise am Opernhaus in Sydney und am Centre Pompidou beteiligt, wo ich erstmals auf ihn traf.
Eines Tages rief er mich in Südfrankreich an und sagte: »Ich steige gerade in das Team von Adrien Fainsilber ein, der den Auftrag für das neue Wissenschafts- und Technikmuseum in Paris erhalten hat. Er möchte mit Glas arbeiten, wie es Norman Foster tut, und ich sagte ihm, dass Sie der Glass Man sind.«
So kamen wir zusammen und gründeten mit RFR ein Ingenieurbüro, das an extrem prestigeträchtigen Bauten wie dem Flughafen Charles de Gaulle oder der Louvre-Pyramide beteiligt war. Ich zog mich aus dem Tagesgeschäft zurück, als ich mit Eco begann, kam aber zurück, als er an einem Gehirntumor erkrankte und 1992 verstarb. Ich blieb dann bei RFR, bis ich meinen 50-prozentigen Anteil an der Firma auf seinen Sohn übertragen hatte.

Peter was a genius, considered by many, to be the best engineer of the 20th century. Among other things, he was the site engineer for the Sydney Opera House, the engineer for the Pompidou Centre in Paris, where I first met him, and a was mentor for many leading architects of the time.
One day he called me in the south of France and said, "I have been asked to join the team of Adrien Fainsilber who has won the competition for the new Museum of Science and Technology in Paris and he wants to do glass like Norman Foster does, so I told him that you did glass for Foster." As a result, we got together and founded RFR, an engineering firm to do the project.
Subsequently we were involved in many extremely prestigious buildings such as the Charles de Gaulle Airport and notably the Louvre Pyramid. When the Eco project started, I retired from the day-to-day business but came back again as managing director when he got brain cancer. After he died in 1992, he left me half the shares and I managed the transfer to his son who eventually took it over.

Emilio Azcárraga

Ich hatte noch nie eine Motoryacht umgesetzt, und Azcárraga gab mir gleich den Auftrag für die revolutionäre Eco. Das war ein unglaublicher Vertrauensbeweis, zumal ich quasi in Personalunion noch das Projektmanagement-Team war, die Werft sowie die Consultants bezahlte und das Budget verwaltete. Zu Beginn der Bauzeit begründete Azcárraga diesen Schritt sehr simpel: »Ich möchte keine ganze Armee beschäftigen. Es ist dein Projekt. Schick mir einfach deine Bankdaten.« Emilio, der mit der Zeit ein guter Freund wurde (Gott hab ihn selig), überwies mir fünf Millionen Dollar und sagte: »Gib Bescheid, wenn du mehr benötigst.« Nach der Ablieferung von Eco bekam ich von Emilio jedes Jahr einen neuen Auftrag, unter anderem für ein schwimmendes Hotel oder etwa eine Überdachung des Fußballstadions in Mexiko-Stadt. Er liebte Eco übrigens so sehr, dass er – an Krebs erkrankt – seine letzten Monate an Bord verbrachte und dort verstarb. Obwohl Eco zunächst nie in Zeitschriften veröffentlicht werden durfte und damit eher wenigen Insidern bekannt war, festigte die Yacht meinen Ruf in der Branche.

I had never built a motor yacht before when in 1986 Azcárraga gave me the commission to design what became the revolutionary Eco. It was an incredible gesture of confidence, especially as I was responsible for everything (that is now done by a big team), owners rep, project management, the budget, paying the shipyard and the consultants etc. When we started, Emilio said, "I don't want to take on the whole German army, it is your project, give me your bank account details". He sent me five million dollars and said: "Let me know when you need more." After the delivery of Eco, Emilio gave me a commission every year, including a floating hotel or a roof over the football stadium in Mexico City. A real patron who loved Eco so much that, suffering from cancer, he spent his final months living and finally and died on board. Although he never allowed the boat to be published it established my reputation in the business.

▸

Larry Ellison

Als Emilio Azcárraga auf der Eco starb, kam die Yacht auf den Brokerage-Markt, wo sie eine ganze Weile gelistet war. Irgendwann kaufte dann Larry Ellison die Yacht, der Gründer von Oracle und inzwischen einer der reichsten Menschen der Welt. Larry beauftragte mich als ursprünglichen Designer mit dem Refit der Yacht, die er in KATANA umbenannte; unter anderem senkten wir das Heck des Eignerdecks ab und installierten einen Basketballkorb auf dem Achterdeck, wo zuvor das Wasserflugzeug parkte. Er bestellte bei mir auch den Entwurf für eine wesentlich größere Yacht, verlor aber das Vertrauen in mich, als er sich das Refit von KATANA anschaute. Eigentlich hatte er mir versprochen, es nicht zu tun – »Sei beruhigt, ich schaue mir keine Operation am offenen Herzen an« – tat es dann aber doch und war schockiert. Den Prozess des Yachtbaus oder -refits kannte er einfach nicht. Das 126-Meter-Projekt SULTAN konnte ich für ihn daraufhin nicht realisieren, es bleibt indes einer meiner liebsten nicht gebauten Entwürfe.

After Emilio Azcárraga died, Eco entered the brokerage market, where it was listed for quite a while. Finally, Larry Ellison, the founder of Oracle and one of the richest people in the world, acquired the yacht. Having been advised to talk to the original designer, Larry hired me to remodel the yacht which he renamed KATANA; among other things, we lowered the owners deck aft and, in place of the seaplane, installed a basketball net on the aft deck. In parallel he asked me for a proposal for a larger yacht however, having been his "new best friend" when we were designing the refit, he lost confidence when, having previously said to me "Don't worry, I don't watch open heart surgery, I won't go to the yard" – he made an impromptu visit to the shipyard shortly before delivery. He was not used to the process and was horrified by the state of his beautiful yacht. My proposal, a radical 126 metre GT powered yacht subsequently known as SULTAN remains one of my favourite unbuilt projects.

Steve Jobs

Als ich mein Büro in London hatte, rief meine Assistentin irgendwann: »Martin, ich habe einen Mister Jobs am Telefon.« Ich ließ ihn durchstellen und fragte: »Sind Sie der Steve Jobs?« Er ließ sich nicht durch sein Vorzimmer durchstellen oder den Kontakt über jemand anderen herstellen, er rief einfach selbst an. Wie sich relativ schnell herausstellte, hatte er nach einem ersten Krebsleiden einige Tage auf KATANA verbracht, die ich ja ursprünglich als ECO gezeichnet und dann für Larry Ellison umgestaltet hatte. Steve wollte mehr über Yachten herausfinden, konnte durch seine Krankheit nicht reisen und lud mich deshalb nach Kalifornien ein, um über ein mögliches Yachtprojekt zu diskutieren. In Palo Alto, wo er in einem recht normalen Haus, umgeben von Apfelbäumen, lebte, nahm er mich sehr herzlich in Empfang. Wir gingen in den Garten, hatten Lunch und diskutierten stundenlang. Am folgenden Tag war die Klingel kaputt, und es gab kein Security-Personal. So ging ich einfach einfach in den Garten, wo niemand war und dann ins (unverschlossene) Haus. Ich hörte Stimmen, und Steve rief: »Martin, wenn du es bist, mach es dir einfach bequem. Ich bin gleich unten.« Wir unterhielten uns erneut für Stunden, gingen spazieren, das Yachtbusiness konnte Steve indes nicht begreifen. Er sagte: »Als ich zu Apple, ohne Gehalt, zurückkam, stellte mir die Firma ein Flugzeug zur Verfügung. Ich durfte allerdings nicht die Motoren oder die Flügel aussuchen, nur die Türdrücker und die Paneele. Wie kann es sein, dass Yachtbau auf einem weißen Blatt Papier beginnt?«

Ich kehrte nach London zurück und präsentierte ihm etwas später Alternativen zu KATANA. Er fragte, warum ich sie überhaupt verändert hätte, und war damit ein typisches Beispiel für so einige Kunden. Sie wissen nicht genau, was sie möchten, bis sie es in der Realität sehen oder betreten. Das Projekt schlief dann ein, als sich Steve für den Launch des ersten iPhones vorbereitete. Etwas enttäuscht war ich schon, dass ich letztlich nicht seine Yacht gestaltete, doch diese drei Treffen mit Steve Jobs gehören zu inspirierendsten, die ich je hatte.

When I had an office in London, my then (not very worldly) receptionist called and said, "Martin, I have a Mr Jobs on the phone, will you take the call?" She put him through and I said, "Is it you Steve Jobs?" He had placed the call himself, no PA or secretary. As it turned out he had spent a lot of time on KATANA lent to him by Larry Ellison and, recovering from his first bout of cancer, wanted to find out more about yachts.

Apologising for not being able to travel because of his illness, Steve invited me to California to discuss find out more. I visited him in his home in Palo Alto, where he lived in a very simple house surrounded by apple trees. Steve was very welcoming, and we sat in the garden discussing yachts and having lunch. When I came back the following day, I tried ringing the bell that was broken (there was no security) no reply, I went into the garden, nobody there, on into the open door of the house where I heard voices upstairs; Steve called down, "Is that you Martin? make yourself at home, I'll be down in a minute." We talked for hours, walked round the block a few times, he said he couldn't understand the yacht business, "When I returned to Apple, without a salary, they gave me an aeroplane, I couldn't choose the engines, the wings etc. just the door handles and the veneer, how come the yacht builders start from a blank sheet of paper each time?"

I returned to London and after making a series of discussion documents went back to see him in Palo Alto. Every time I proposed something different from KATANA, he said "why change it?" and what became apparent was that, in common with many clients, he didn't know what he wanted until he saw it, then the project went to sleep for a long time while he was preparing for the launch of the first iPhone. Although in the end I didn't design his yacht, those three intimate meetings with Steve Jobs in his home where among the most inspiring I've ever had.

►

Philippe Starck

Der Broker Nick Edmiston machte mich mit Andrei Melnitschenko bekannt, einem sehr reichen Russen, der eine sehr große Yacht bauen wollte. Das erste Treffen dazu fand in einer angemieteten Villa auf Cap Ferrat statt. Ich nahm ein Modell meines Entwurfs FD100 und erläuterte das Styling vor dem Auftraggeber und einer Gruppe. Die FD100 ist die kleinere Version meines Projekts SULTAN, und es geschah das erste Mal in meiner Karriere, dass ich bei einem solchen Pitch Applaus erhielt. Die ganze Gruppe klatschte, und klatschte und ich wurde zur nächsten Präsentationsrunde nach Barcelona eingeladen. Die PowerPoint-Folien waren total stimmig, nur die Lichtsituation in dem spanischen Hotel war eine Katastrophe. Es war viel zu hell, sodass man den Inhalt der Folien nicht gut erkennen konnte. Ein Kollege von mir, der seine Ideen per Hand auf Papier skizzierte, bekam letztlich den Auftrag von Melnitschenko. Es war etwas Pech und auch noch blöd gelaufen.

Auf der folgenden Monaco Yacht Show traf ich den Kollegen, der das Rennen gemacht hatte. Zu meiner Verwunderung berichtete er, dass das Projekt zum Stillstand gekommen sei. Er hätte seit dem Event in Barcelona eigentlich nichts mehr gehört. Wenige Wochen später erhielt ich allerdings einen etwas ominösen Anruf aus Zypern. Es ging erneut um das Melnitschenko-Projekt, das nun von Philippe Starck betreut wurde, der für Melnitschenko auch gerade ein Apartment gestaltete. Der Anrufer sollte den Kontakt zu mir herstellen, da Starck Hilfe bei der Konstruktion benötigte; Melnitschenko selbst hatte die Zusammenarbeit empfohlen. Ich traf mich also mit meinem Kollegen aus Paris, und wir einigten uns darauf, dass er die Verantwortung für das Konzept und das Interieur bekommen sollte, ich war für die Technik und die Konstruktion zuständig. Zusammen entwickelten wir das Projekt SF99 – S für Starck, F für Francis – aus dem später die Motoryacht A werden sollte. Nachdem die sehr außergewöhnliche Yacht gewassert war, vergaß Starck zwar mitunter meinen Beitrag zum Gelingen, trotzdem muss ich sagen: Philippe ist ein grandioser Kollege, mit dem die Zusammenarbeit sehr inspirierend war, indem ich (auf meine alten Tage) etwa lernte, wie er sich als Marke inszenierte.

Yacht broker Nick Edmiston introduced me to Andrey Melnichenko, a very wealthy Russian who wanted to build a large yacht. The first meeting took place in a rented villa on Cap Ferrat. I took a model of my FD100 project (a smaller version of my SULTAN project) as an example of my work in addition to a Power Point presentation made to Andrey and a group of his friends. After which the whole group clapped, it was the first time in my career that I got applause when making a pitch.

I was then invited to prepare a proposal for the next round of presentations in Barcelona. A PowerPoint presentation had worked before but, this time, the lighting situation in the Spanish hotel was much too bright, so that my proposal could not be seen properly, a disaster.

The same day another designer sat on the floor and sketched his ideas on paper and apparently got the commission.

However, at the following Monaco Yacht Show I met the other designer who to my astonishment said that the project had come to a standstill, he hadn't heard anything since the event in Barcelona.

Coincidentally the same day, following a rather strange phone call from Cyprus, I was interviewed by someone who said that he represented Melnichenko and that Starck, for whom he was doing an apartment, had come up with a concept but not being a naval architect, didn't know how to realise it therefore Melnichenko had suggested we work together.

After some preliminary negotiations on IP rights, I finally met with Philippe, got on well, and we agreed that he was the chief concept and interior designer and I was the chief technical and naval designer, the yard name was SF 99 for Starck and Francis. Unfortunately, after the yacht subsequently called (motor yacht) A, was launched with much acclaim, Philippe wasn't happy to share the credit.

However, I have to say I greatly enjoyed the man and our collaboration and, in the process, observed with great interest how to build a successful brand.

Um den Entwurf für die Erweiterung des Louvre gab es heftige öffentliche Diskussionen. Das Museum war das wichtigste Projekt unter den zahlreichen Bauvorhaben des französischen Präsidenten François Mitterrand, der I. M. Pei persönlich mit dem Auftrag betraute. Der legendäre chinesisch-amerikanische Stararchitekt schlug vor, den Haupteingang im Innenhof des Louvre zu platzieren, und zwar als gewaltige Glaspyramide. Ich rutschte in diese Planung durch meine Glaskenntnisse, für die ich in der Architekturszene recht bekannt bin. Über 80 Prozent der Pariser lehnten den Entwurf allerdings ab, und auch Jacques Chirac, damals Bürgermeister von Paris, musste erst restlos überzeugt werden. Ich schlug Pei deshalb vor, die Größe der Pyramide originalgetreu zu simulieren. Mein Plan sah einen gewaltigen Kran vor, von dem Seile herabhingen, die die Seiten der Pyramide simulierten. Pei fand das großartig, und so parkte irgendwann dieser Kran im Innenhof, und Chirac gab den Startschuss für eines der bekanntesten Bauwerke der Welt.

Monate später rief mich Peis lokaler Architekt an, um zu fragen, wie wir Segler eigentlich die Masten unserer oder anderer Boote designen. Es ging natürlich immer noch um die Konstruktion des Louvre. Ich traf mich mit Pei, erklärte ihm die Vorzüge des Rigging von Segelyachten und empfahl ihm Tim Eliasson von Navtec. Ergebnis: Eliasson bekam den Job, gründete die Firma TriPyramid Structures und ist nun einer der Weltmarktführer für die Konstruktion von Glasstrukturen – unter anderem ist er an allen Apple-Stores beteiligt.

The design for the extension of the Louvre was the subject of heated public debate. The museum was the most important project among the numerous cultural projects of the French President François Mitterrand, who had personally entrusted I. M. Pei with the task. The legendary Chinese American star architect proposed that the main entrance to the museum should be in the form of a large glass pyramid in the centre of the courtyard. My first involvement came in the form of a request to find a way of simulating the pyramid to demonstrate to Jacques Chirac, then the mayor of Paris, the validity of the idea. I came up with the idea of using a large tower crane to suspend, via an almost invisible thin steel wire, a net of foam-covered cables outlining the main structure of the pyramid. This was a great success because it only needed to be erected during Mr Chirac's visit and he approved it, so the project went ahead. Several months later Pei's local architect called me to ask me how I designed the masts of my sail boats. They had some Canadian engineers who had proposed a cable stayed mast like structure and didn't know how to make it work. I had a meeting with Pei at which I pointed out that in boats we would do it a different way using discontinuous rod rigging instead of the wire he had chosen a year earlier. He agreed so I phoned Tim Eliasson then at Navtec and told him I thought I had a project for him. The result was, he set up a new company called TriPyramid Structures they got the job and are now world leaders in glass structures such and the Apple stores. Subsequently RFR became Pei's engineers of choice doing the inverted Pyramid and then the Luxemburg museum for him.

I. M. Pei

▶

Frank Stella

Frank ist für mich einer der größten Künstler unserer Zeit. Er arbeitete lange mit Peter Rice zusammen, bis Peter krank wurde. Ich sprang ein und realisierte mit und für ihn eine große Skulptur für ein Hotel in Singapur. Die französischen Ingenieure, die seine Projekte sonst realisierten, umging ich und nutzte Bootsbauer aus Antibes, die einen super Job erledigten. Nachdem Frank verstärkt mit Skulpturen arbeitete, kooperierten wir eng und freundeten uns an. Ich half ihm, mit meinen Bootsbau-, Konstruktions-, 3-D-Druck- und Computerkenntnissen diese manchmal unglaublich komplizierten und großen Kunstwerke zu realisieren – die größte, in einer Werft gebaute, wiegt immerhin stattliche 45 Tonnen. Frank, der inzwischen Veronika Schmid, eine meiner ehemaligen Mitarbeiterinnen beschäftigt, ist eine Inspiration für mich, und ich bin glücklich und dankbar, ihn als Kollegen und Freund bezeichnen zu können.

Frank is for me one of the greatest artists of our time, he had been working with Peter Rice for several years before Peter became ill when I took over firstly to build a large piece of sculpture destined for a hotel in Singapore.

I abandoned the engineers in Paris and built it with artisan boat builders in Antibes thus starting a long collaboration and friendship. After working with bigger and bigger sculptures, one of which was built in a shipyard and weighed 45 tons, I introduced him to parametric computer modelling and 3D printing and he now employs Veronika Schmid, one of my ex staff, to work on his numerous projects. Frank continues to be an inspiration to me and is above all a good friend.

Rex Harbour

Rex ist ein sehr zurückgezogener, aber loyaler Kunde von mir, inzwischen auch ein Freund, dem ich zeitweilig bei der Visualisierung von Architekturprojekten half. Vor rund zehn Jahren fragte er mich dann, ob ich ihm beim Umbau einer Villa helfen könne, die er in meiner unmittelbaren Nachbarschaft in Südfrankreich gekauft hatte. Das geschah zu einer Zeit, als ich gerade nicht sehr viele Aufträge hatte, und seine Anfrage brachte mich zurück ins Architekturgeschehen. Das Projekt wuchs um einen spektakulären Glaspool und einen »Wassergarten« – und unsere Freundschaft wuchs mit. Unsere gemeinsamen Interessen – unter anderem Architektur und Musik – diskutieren wir oft an lauen Sommerabenden in seinem wundervollen Garten.

Rex is a very discreet but loyal client and above all friend who I had assisted from time to time with the visualization of architectural projects. About ten years ago he asked me if I would help him in converting the villa he had recently acquired near to my home in the South of France. His project came at time when I didn't have much work and his encouragement got me back into architecture. The project grew to include a glass pool, water garden and dependences and our friendship with it. Sharing common interests in architecture, music and many other things we often enjoy summer evenings in the wonderful garden he has since created.

Paul Allen

Paul Allen lernte ich über seinen sogenannten Marinemanager kennen. David Reams gefiel meine Arbeit, und zusammen mit mehreren anderen Designern lud er mich deshalb ein, um einen Vorschlag für ein neues Schiff als Ersatz für Paul Allens OCTOPUS zu unterbreiten. Nach der Präsentation zahlreicher Skizzen schlug ich vor, dass es sehr nützlich wäre, sich mit Allen zu treffen, um zu sehen, ob wir überhaupt einen Draht zueinander finden würden. Eine Yacht für ihn zu entwerfen, wenn er den Designer nicht mochte, wäre extrem schwierig.

David antwortete, dass er während der vergangenen zwei Jahre, für die er für Allen gearbeitet habe, kein Treffen erlebt hätte, das länger als 20 Minuten gedauert hätte. Er würde es aber versuchen und Allen ein Treffen vorschlagen. Dessen Antwort lautete: »Warum sollte ich den Designer treffen?« David entgegnete daraufhin: »Weil es dein Boot sein wird, nicht meines.« Eines Sonntagabends erhielt ich also einen Anruf von David, der fragte, ob ich am folgenden Nachmittag in London sein könnte, um Allen zu treffen. Am nächsten Morgen stieg ich in das erste Flugzeug und kam rechtzeitig in seinem Londoner Haus an, um mich in Ruhe für eine Präsentation vorzubereiten.

Als Allen erschien, setzten wir uns hin und begannen zu reden: über Yachten, seine und einige meiner Projekte, die er kannte. Das Treffen dauerte etwa 45 Minuten, ohne dass ich ihm meine Präsentation zeigte. Ich hatte immer davon geträumt, den Kunden zuerst kennenzulernen und dann weiterzumachen. Hier schien das zu klappen.

Ich rief nach dem Treffen sofort Dave an, der sehr verblüfft war. Jedenfalls bekam ich sehr schnell die Zusage, dass ich für das Projekt engagiert werden würde – und so begann eine lange Zusammenarbeit.

Nach mehrmonatiger Entwicklung mit Allens Team und Spezialisten für Hubschrauber, Unterwasserforschung bis hin zu Aufnahmestudios wurde aus unser aller Know-how ein 158 Meter langes Schiff, das wir in eine Werftausschreibung gaben. Nach vielen weiteren Monaten der technischen Entwicklung, Entwicklungsverbesserungen und Kostensenkungsmaßnahmen hatten wir einen endgültigen Vorschlag, der alle Kriterien erfüllte. David konsultierte Allen, der nach reiflicher Überlegung sagte: »Für dieses Geld könnte ich mein Institut für Hirnforschung 17 Jahre lang finanzieren, und ich finde es schwer zu rechtfertigen, das Geld für mich selbst auszugeben.«

So sehr ich es auch bedauerte, dass ich das größte Projekt meiner Karriere nicht realisieren konnte, ich muss Pauls Motive bewundern. Solche Menschen wie ihn und noch mehr davon benötigt diese Welt. Leider verstarb er kürzlich. Nicht nur ich vermisse ihn schmerzlich.

I got to meet Paul Allen through his marine manager David Reams who although we had never met liked my work; along with several other designers he invited me to make a proposal for a new vessel to replace OCTOPUS. After presenting numerous sketches I suggested that it would be useful to meet with the owner to see if we got on, it would be difficult to design a boat for him if he didn't like the designer.

Dave replied that during the two years he had worked for Paul, the longest meeting he had was about twenty minutes and he doubted if Paul would have the time. However, he suggested a meeting to Paul who replied, "why would I need to meet the designer?" to which Dave replied, "because it will be your boat not mine." So, one Sunday night, I got a call from Dave to ask me if I could be in London the following afternoon to meet Paul, of course, I jumped on the first plane next morning and got to his London home in time to set up for a slide presentation.

When Paul arrived, we sat down and started talking about yachts, his and some of my projects that he knew of. The meeting lasted about 45 minutes, but we never got around to looking at my presentation, something I had always dreamed of, getting to know the client first. I immediately called Dave who was amazed and later called me back to say that they would engage me to work up the project in collaboration with the Vulcan team, thus started a long collaboration.

Following many months of development with the numerous Vulcan specialists in everything from helicopters, underwater research to recording studios, we arrived at a 158 metre vessel and went out to bid to three yards. After many more months of technical development refinement and cost cutting exercises, we got a final proposal that ticked all the owner's requirement boxes. David presented this all to Paul who after much reflection said, "For that money, I could fund my institute for brain research for 17 years and I find it difficult to justify spending the money on myself."

Much as I regretted not building the biggest project of my career, I must admire Paul's motives, the world needs more people like him. Very sadly as it has turned out, Paul died before the project would have been finished, he will be greatly missed.

James O'Callaghan

James und sein Unternehmen Eckersley O'Callaghan sind – meiner Meinung nach – heute die führenden Ingenieure weltweit und würdige Nachfolger meines verstorbenen Freundes und Partners Peter Rice. Ich lernte James aufgrund unseres gemeinsamen Interesses an der Glastechnik kennen, seiner engen Zusammenarbeit mit Steve Jobs und der damit verbundenen Mitarbeit an allen Apple Stores; später weitete sich James' Engagement auch auf Yachten aus. Er ist ein extrem inspirierender Ingenieur, der für die größte Entwicklung in der Strukturglastechnologie seit der Erfindung des Floatglases durch Pilkington im Jahr 1952 verantwortlich war. James ist zu einem wertvollen Freund geworden, den ich bei meinen Projekten, seien es Gebäude, Schiffe oder Yachten, zurate ziehe, wann immer ich kann. Gerade wurde er mit der Goldmedaille der Institution of Structural Engineers ausgezeichnet.

James and his firm Eckersley O'Callaghan are the leading engineers today, worthy successors to my late friend and partner Peter Rice. I got to know James because of our common interest in glass engineering and his close involvement with Steve Jobs, all the Apple stores and subsequently in yachts. He is the inspirational engineer who was responsible for the biggest development in structural glass technology since the invention of float glass by Pilkington in 1952. James has become a valuable friend who I consult whenever I can on my projects, buildings, ships or yachts, he has just been awarded the 2019 Gold Medal of the Institution of Structural Engineers, the professions highest honour.

Mark Adams

Ein alter Freund, den ich seit 30 Jahren nicht mehr gesehen hatte, rief mich eines Tages aus heiterem Himmel an, um mir zu sagen, dass er mich gern mit Mark Adams vernetzen würde, dem CEO von Vitsœ und Hersteller von Dieter Rams Möbelsystemen.

Adams hatte mehrere Architekten und Ingenieure auf der Suche nach jemandem für den Entwurf seines neuen Hauptsitzes befragt, aber nach seinen eigenen Worten hatte er »sich in seinem Leben noch nie so verachtet gefühlt« und suchte nun eine Alternative. Das Briefing listete unter anderem: einen Holzrahmen, Paxtons Crystal Palace, Yachten der J-Klasse und ein frühes Foster-Gebäude. Paxton war schon lange ein Idol und beliebtes Thema bei meinen Vorlesungen vor Architekturstudenten gewesen, er war weder Architekt noch Ingenieur, sondern ein Gärtner, der 1851 den Crystal Palace in elf Monaten baute – heute wäre so etwas natürlich völlig unmöglich.

Eines Samstagmorgens stellte ich jedenfalls Adams und seiner Frau und Partnerin Jenny Moncur eine Präsentation vor, und es zeigte sich bald, dass wir auf einer Wellenlänge waren. Am darauffolgenden Montag begann ich eine Peer-Review der bestehenden Projekte, und wir kamen schnell zu dem Schluss, dass es notwendig war, neu zu beginnen.

So begann eine lange und fruchtbare Zusammenarbeit, die dazu geführt hat, dass die Vitsœ-Zentrale in Leamington Spa ein bemerkenswert unauffälliges und umweltfreundliches Gebäude wurde, das neue Maßstäbe setzt. Dabei hat es einen neuen Karriereweg für Adams und mich eröffnet – die Konstruktion und Produktion von Holzsystemhäusern.

An old friend who I had not seen for 30 years called me out of the blue one day to say that he thought he might have a job for me and set up for me to meet Mark Adams, he the owner of a company called Vitsœ who make Dieter Rams furniture systems.

He had been through several architects and engineers in the search for someone to design his new headquarters but in his own words "had never felt so despised in his life" and so was seeking someone else. Among other things the brief referred to: a timber frame, Paxton's Crystal Palace, J class yachts and an early Foster building.

Paxton had long been an idol and favourite topic of mine when giving lectures to architectural students, he was neither architect or engineer but a gardener who in 1851 built the Crystal Palace in eleven months something quite impossible today.

One Saturday morning I made a presentation to Mark and his wife and partner Jenny Moncur and it was soon apparent that we were on the same page, the following Monday I started a peer review of the existing projects and we quickly concluded the need to start again.

So started a long and fruitful collaboration which has resulted in the Vitsœ headquarters in Leamington Spa a remarkably understated, environmentally friendly building that sets new standards. In the process it has opened a new career path for Mark and me, the design and production of timber system buildings.

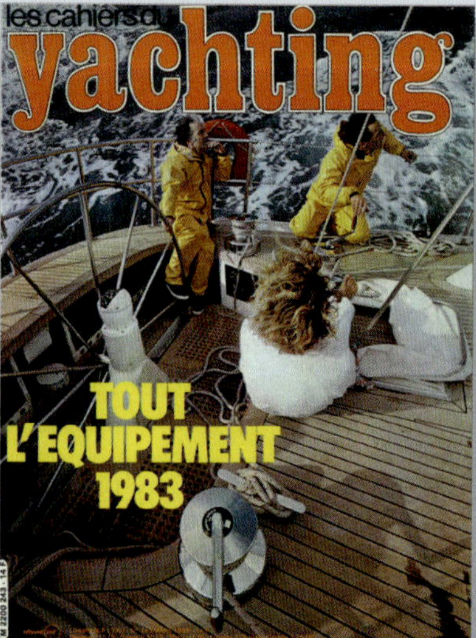

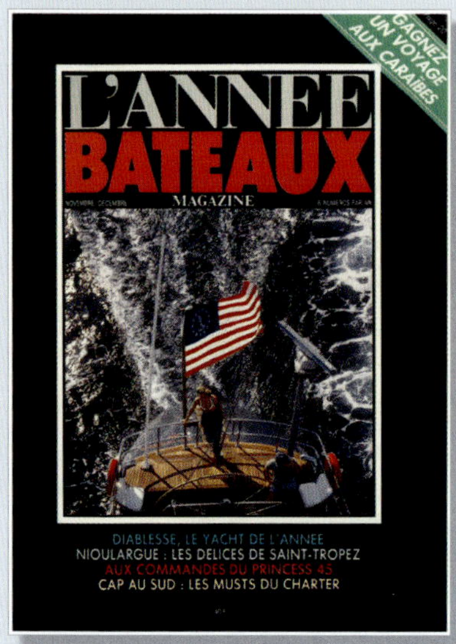

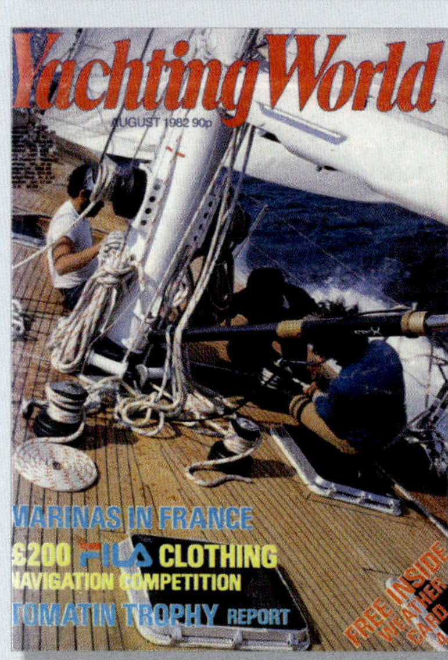

Titel Covers

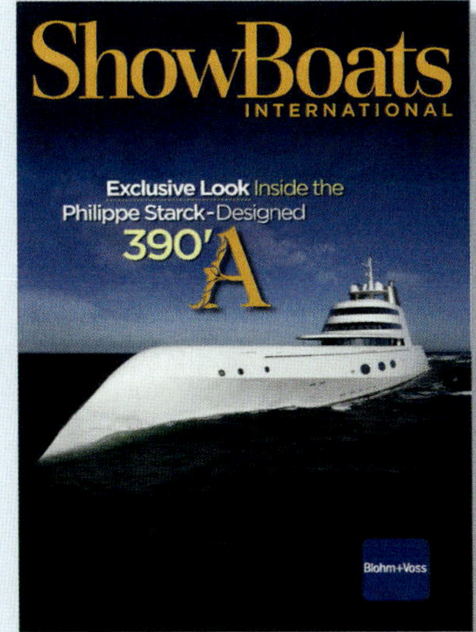

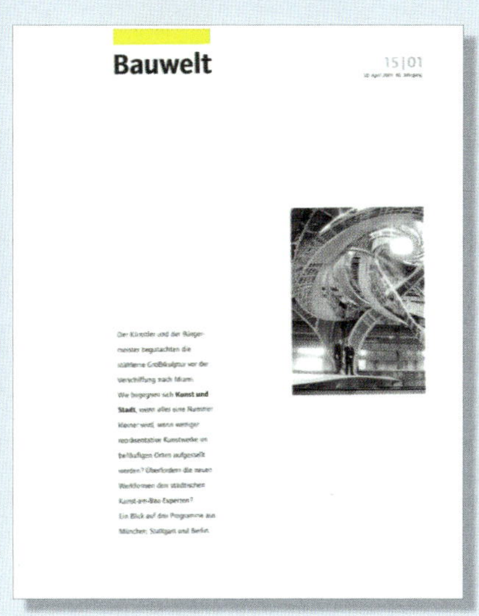

TITEL, TITEL, TITEL

Die Welt der Superyachten ist eigentlich eine verschlossene Szene: Niemand soll wissen, wer sich gerade 60 Meter gebraucht gekauft, 80 Meter gechartert oder 100 Meter in Auftrag gegeben hat. Und doch gibt es weltweit eine Medienszene, die sich nur mit diesem Thema beschäftigt. Martin Francis war indes nie scheu, mit der Presse zu sprechen. Die Journalisten, die seine Grenzen respektieren – »this is strictly confidential« –, die respektiert er ebenfalls und arbeitet mit ihnen zusammen, wenn denn eines seiner Projekte veröffentlicht werden darf. Und da diese alles andere als langweilig sind, könnte Francis mit den Zeitschriften, auf denen seine Entwürfe zu sehen sind, ganze Wände tapezieren.

COVERS, COVERS, COVERS

The world of superyachts is actually a closed scene: Nobody should know who just bought 60 metres used, chartered 80 metres or commissioned 100 metres. And yet there is a worldwide media scene with magazines, websites and social media influencers that only deals with this topic. Martin Francis was never shy to talk to the press. He also respects the journalists who respect his limits – "this is strictly confidential" – and works with them when one of his projects is allowed to be published. And since these are anything but boring, Francis could wallpaper entire walls in his home with the magazines on which his designs have been published.

Titel
Covers

Danke
Thank you

VIELEN DANK AN

Peter Tamm, der dieses Buch ermöglicht hat und mir angeboten hat, mein Archiv zu pflegen.
Marcus Krall für die Zusammenstellung und das Schreiben des Textes.
Marisa Tippe für die Gestaltung des Buches und die Bewältigung aller meiner Änderungsvorschläge.

Für ihre Liebe und Unterstützung, in guten und schlechten Zeiten, Dank an:
Meine Frau Sandrine Melot.
Meine Kinder Tania, Kim, Annina und Melissa.
Meine Enkelkinder Ruben, Scarlett, Nina und Rosa.
Meine Stiefkinder Ambroise und Leonore.
Meinen Bruder Simon und seine Frau Pam für alles, was sie für Mel getan haben.

THANKS, ACKNOWLEDGEMENTS

Thanks to Peter Tamm for making this book possible and for offering to look after my archive.
Marcus Krall for getting it together and writing the text.
Marisa Tippe for designing the book and putting up with all my suggested changes.

For their love and support, through good times and bad, thanks to:
My wife Sandrine Melot.
My children Tania, Kim, Annina and Melissa.
My grandchildren Reuben, Scarlett, Nina and Rosa.
My stepchildren Ambroise and Leonore.
My brother Simon and his wife Pam for all that they have done for Mel.

■

BILDNACHWEIS/PHOTO CREDITS

FOTOGRAF/PHOTOGRAPHER SEITE/PAGE

Bel Geddes Foundation 19
Celebrity Cruises 136, 137, 140, 141
Colahan, Mirelle 13, 75
Denton, Pat 8, 78, 142
FleetMon 21
Foster + Partners 111, 124, 125, 127, 145
Francis, Martin . . . . Cover, 11, 14, 20, 32, 33, 34, 35, 36, 44, 48, 49, 57, 67, 74, 75, 77, 113, 116, 117, 120, 121, 126, 129, 132, 133, 134, 135, 147, 152
Francis, Hugh 10
Lindner, Dirk 111, 112, 113, 115
Living Oceans Foundation . . 38, 39, 41
Lousada, Sandra 16, 130, 131
Marimon, Toni 40
Melot, Sandrine 20, 22, 108, 144
Norman Foster Foundation . . 21
Orlewicz, Victor 151
Plisson, Philippe 25
Plisson, Guillaume . . . . Cover, 14, 17, 24, 26, 27, 29, 30, 31, 37, 46, 47, 50, 51, 52, 53, 54, 55, 56
Poli, Tomasso 72, 73
Rabinowitz, Neil 58, 59, 60, 61, 62, 63, 64, 65
Ratcliff, Justin 10
Russel, Ethan 12
Ryckaert, Marc 138, 139
Seyfferth, Peter 24, 42, 43
Sezerat, Antoine 25
Silver Arrows Marine 66, 68, 69, 70, 71
Spath, Dagmar 19
Stehnken, Andreas 15
Verdure, Michel 137

Nicht bei allen Abbildungen konnten die Inhaber der Bildrechte ermittelt werden. Der Verlag bittet freundlich um Kontaktaufnahme: Maximilian Verlag, Stadthausbrücke 4, 20355 Hamburg.

The owners of the copyrights could not be determined for all images. The publisher kindly asks you to contact us: Maximilian Verlag, Stadthausbrücke 4, 20355 Hamburg.

IMPRESSUM

Ein Gesamtverzeichnis der lieferbaren Titel schicken wir Ihnen gerne zu. Bitte senden Sie eine E-Mail mit Ihrer Adresse an vertrieb@koehler-books.de
Sie finden uns auch im Internet unter www.koehler-books.de

Bibliografische Information der Deutschen Nationalbibliothek
Die Deutsche Nationalbibliothek verzeichnet diese Publikation in der Deutschen Nationalbibliografie; detaillierte bibliografische Daten sind im Internet über http://dnb.d-nb.de abrufbar.

ISBN 978-3-7822-1327-1

Gestaltung: Marisa Tippe, Hamburg
Druck: Firmengruppe APPL, aprinta Druck, Wemding

IMPRINT

We will gladly send you a complete list of available titles. Please send an e-mail with your address to: vertrieb@koehler-books.de
Our web address is: www.koehler-books.de

Bibliographical information of the German National Library
The German National Library lists the original German publication in the German National Bibliography; detailed bibliographical data is available in the internet via http://dnb.d-nb.de.

ISBN 978-3-7822-1327-1

Design: Marisa Tippe, Hamburg
Printed by: Firmengruppe APPL, aprinta Druck, Wemding

■